## WORKS OF C. E. GREENE

PUBLISHED BY

## JOHN WILEY & SONS.

**Graphics for Engineers, Architects, and Builders.**
A manual for designers, and a text-book for scientific schools.

**Trusses and Arches;**
Analyzed and Discussed by Graphical Methods. In three parts—published separately.

**Part I. Roof Trusses:**
Diagrams for Steady Load, Snow and Wind. 8vo, 80 pp., 3 folding plates. Revised Edition. $1.25.

**Part II. Bridge Trusses:**
Single, Continuous, and Draw Spans; Single and Multiple Systems; Straight and Inclined Chords. 8vo, 190 pp., 10 folding plates. Fifth Edition, Revised. $2.50.

**Part III. Arches,**
In Wood, Iron, and Stone, for Roofs, Bridges, and Wall-openings; Arched Ribs and Braced Arches; Stresses from Wind and Change of Temperature. 8vo, 194 pp., 8 folding plates. Third Edition, Revised. $2.50.

**Structural Mechanics.**
Comprising the Strength and Resistance of Materials and Elements of Structural Design. With Examples and Problems. By the late Charles E. Greene, A.M., C.E. New Edition, Revised and Enlarged by A. E. Greene. 8vo, viii + 244 pages, 99 figures. $2.50 *net*.

# STRUCTURAL MECHANICS

COMPRISING THE

STRENGTH AND RESISTANCE OF MATERIALS AND
ELEMENTS OF STRUCTURAL DESIGN

WITH EXAMPLES AND PROBLEMS

BY

CHARLES E. GREENE, A.M., C.E.
LATE PROFESSOR OF CIVIL ENGINEERING, UNIVERSITY OF MICHIGAN

REVISED BY A. E. GREENE
JUNIOR PROFESSOR OF CIVIL ENGINEERING, UNIVERSITY OF MICHIGAN

*THIRD EDITION, REVISED.*
FIRST THOUSAND

NEW YORK
JOHN WILEY & SONS
LONDON: CHAPMAN & HALL, LIMITED
1907.

# PREFACE.

The author, in teaching for many years the subjects embraced in the following pages, has found it advantageous to take at first but a portion of what is included in the several chapters, and, after a general survey of the field, to return and extend the investigation more in detail. Some of the sections, therefore, are not leaded and can be omitted at first reading. A few of the special investigations may become of interest only when the problems to which they relate occur in actual practice.

It is hoped that this book will be serviceable after the class-room work is concluded, and reference is facilitated by a more compact arrangement of the several matters than the course suggested above would give. The attempt has been made to deal with practicable cases, and the examples for the most part are shaped with that end in view. A full index will enable one to find any desired topic.

The treatment of the subject of internal stress is largely graphical. All the constructions are simple, and the results, besides being useful in themselves, shed much light on various problems. The time devoted to a careful study of the chapter in question will be well expended.

The notation is practically uniform throughout the book, and is that used by several standard authors. Forces and moments are expressed by capital letters, and unit loads and stresses by small letters. The coordinate $x$ is measured along the length of a piece, the coordinate $y$ in the direction of variation of stress in a section, and $z$ is the line of no variation of stress, that is, the line parallel to the moment axis.

One who has mastered the subjects discussed here can use the current formulas, the pocket-book rules, and tables, not blindly, but with discrimination, and ought to be prepared to design intelligently.

# TABLE OF CONTENTS.

|  | PAGE |
|---|---|
| INTRODUCTION | 1 |

### CHAPTER I.
ACTION OF A PIECE UNDER DIRECT FORCE.................................. 6

### CHAPTER II.
MATERIALS.................................................................. 19

### CHAPTER III.
BEAMS..................................................................... 38

### CHAPTER IV.
MOMENTS OF INERTIA OF PLANE AREAS....................................... 71

### CHAPTER V.
TORSION................................................................... 81

### CHAPTER VI.
FLEXURE AND DEFLECTION OF SIMPLE BEAMS................................. 87

### CHAPTER VII.
RESTRAINED BEAMS: CONTINUOUS BEAMS..................................... 107

### CHAPTER VIII.
PIECES UNDER TENSION.................................................... 127

### CHAPTER IX.
COMPRESSION PIECES: COLUMNS, POSTS, AND STRUTS.......................... 137

### CHAPTER X.
SAFE WORKING STRESSES................................................... 153

## CHAPTER XI.
INTERNAL STRESS: CHANGE OF FORM.................................167

## CHAPTER XII.
RIVETS: PINS............................................................192

## CHAPTER XIII.
ENVELOPES: BOILERS, PIPES, DOME....................................203

## CHAPTER XIV.
PLATE GIRDER...........................................................221

## CHAPTER XV.
SPRINGS: PLATES........................................................229

## CHAPTER XVI.
REINFORCED CONCRETE..................................................235

# NOTATION.

$b$, breadth of rectangular beam.
$C$, shearing modulus of elasticity.
$d$, diameter.
$E$, modulus of elasticity, Young's modulus.
$F$, shear in beam.
$f$, unit stress.
$h$, height of rectangular beam.
$I$, rectangular moment of inertia.
$i$, slope of elastic curve.
$J$, polar moment of inertia.
$k$, a numerical coefficient.
$l$, length of member.
$\lambda$, unit change of length.
$M$, bending or resisting moment.
$P$, reaction of beam; load on tie or post.
$p$, unit stress; unit pressure in envelope, Ch. XIII.
$q$, unit shear.
$R$, radius of circle.
$r$, radius of gyration; radius of envelope, Ch. XIII.
$\rho$, radius of curvature.
$S$, area of cross-section.
$T$, torsional moment; stress in envelope, Ch. XIII.
$v$, deflection of beam.
$W$, concentrated load on beam.
$w$, intensity of distributed load on beam.
$y_1$, distance from neutral axis to extreme fibre of beam.
$x, y, z$, coordinates of length, depth, and breadth of beam.

# STRUCTURAL MECHANICS.

## INTRODUCTION.

**1. External Forces.**—The engineer, in designing a new structure, or critically examining one already built, determines from the conditions of the case the actual or probable external forces which the structure is called upon to resist. He may then prepare, either by mathematical calculations or by graphical methods, a sheet which shows the maximum and minimum direct forces of tension and compression which the several pieces or parts of the structure are liable to experience, as well as the bending moments on such parts as are subjected to them.

These forces and moments are determined from the requirements of equilibrium, if the pieces are at rest. For forces acting in one plane, a condition which suffices for the analysis of most cases, it is necessary that, for the structure as a whole, as well as for each piece, there shall be no tendency to move up or down, to move to the right or left, or to rotate. These limitations are usually expressed in Mechanics as, that the sum of the X forces, the sum of the Y forces, and the sum of the moments shall each equal zero.

If the structure is a machine, the forces and moments in action at any time, and their respective magnitudes, call for a consideration of the question of acceleration or retardation of the several parts and the additional maximum forces and moments called into action by the greatest rate of change of motion at any instant. Hence the weight or mass of the moving part or parts is necessarily taken into account.

Finally, noting the rapidity and frequency of the change of force and moment at any section of any piece or connection, the engineer selects, as judgment dictates, the allowable stresses of the several kinds per square inch, making allowance for the effect of impact, shock, and vibration in intensifying their action, and proceeds to find the necessary cross-sections of the parts and the proportions of the connections between them. As all structures are intended to endure the forces and vicissitudes to which they are usually exposed, the allowable unit stresses, expressed in pounds per square inch, must be *safe stresses*.

It is largely with the development of the latter part of this subject, after the *forces* have been found to which the several parts are liable, that this book is concerned.

**2. Ties, Struts, and Beams.**—There are, in general, three kinds of pieces in a frame or structure: ties or tension members; columns, posts, and struts or compression members; and beams, which support a transverse load and are subject to bending and its accompanying shear. A given piece may also be, at the same time, a tie and a beam, or a strut and a beam, and at different times a tie and a strut.

**3. Relation of External Forces to Internal Stresses.**—The forces and moments which a member is called upon to resist, and which may properly be considered as *external* to that member, give rise to actions between all the *particles* of material of which such a member is composed, tending to move adjacent particles from, towards, or by one another, and causing change of form. There result *internal stresses*, or resistances to displacement, between the several particles.

These internal stresses, or briefly stresses, must be of such kind, magnitude, distribution, and direction, at any imaginary *section* of a piece or structure, that their resultant force and moment will satisfy the requirements of equilibrium or change of motion with the external resultant force and moment at that section; and no stress per square inch can, for a correct design, be greater than the material will safely bear. Hence may be

## INTRODUCTION.

determined the necessary area and form of the cross-section at the critical points, when the resultant forces and moments are known.

**4. Internal Stress.**—There are three kinds of stress, or action of adjacent particles one on the other, to which the particles of a body may be subjected, when external forces and its own weight are considered, viz.: *tensile* stress, tending to remove one particle farther from its neighbor, and manifested by an accompanying stretch or elongation of the body; *compressive* stress, tending to make a particle approach its neighbor, and manifested by an accompanying shortening or compression of the body; and *shearing* stress, tending to make a particle move or slide laterally with reference to an adjacent particle, and manifested by an accompanying distortion. Whether the stress produces change of form, or the attempted change of form gives rise to internal stresses as resistances, is of little consequence; the stress between two particles and the change of position of the particles are always associated, and one being given the other must exist.

**5. Tension and Shear, or Compression and Shear.**—If the direction of the stress is oblique, that is, not normal or perpendicular, on any section of a body, the stress may be resolved into a tensile or compressive stress normal to that section, and a tangential stress along the section, which, from its tendency to cause sliding of one portion of the body by or along the section, has been given the name of shear, from the resemblance to the action of a pair of shears, one blade passing by the other along the opposite sides of the plane of section. Draw two oblique and directly opposed arrows, one on either side of a straight line representing the trace of a sectional plane, decompose those oblique stresses normally and tangentially to the plane, and notice the resulting directly opposed tension or compression, and the shear. Hence tension and shear, or compression and shear, may be found on any given plane in a body, but tension and compression cannot simultaneously occur at one point in a given area.

**6. Sign of Stress.**—Ties are usually slender members; struts have larger lateral dimensions. Longitudinal tension tends to diminish the cross-section of the piece which carries it, and hence may conveniently be represented by —, the negative sign; longitudinal compression tends to increase the cross-sectional area and may be called + or positive. Shear, being at right angles to the tension and compression in the preceding illustration, has no sign; and lies, in significance, between tension and compression. If a rectangular plate is pulled in the direction of two of its opposite sides and compressed in the direction of its other two sides, there will be some shearing stress on every plane of section except those parallel to the sides, and nothing but shear on two certain oblique planes, as will be seen later.

**7. Unit Stresses.**—These internal stresses are measured by units of pounds and inches by English and American engineers, and are stated as so many pounds of tension, compression, or shear per square inch, called unit tension, compression, or shear. Thus, in a bar of four square inches cross-section, under a total pull of 36,000 pounds centrally applied, the internal unit tension is 9,000 pounds per square inch, provided the pull is uniformly distributed on the particles adjacent to any cross-section. If the pull is not central or the stress not uniformly distributed, the average or mean unit tensile stress is still 9,000 pounds.

If an oblique section of the same bar is made, the total *force* acting on the particles adjacent to the section is the same as before, but the area of section is increased; hence the unit stress, found by dividing the force by the new area, is diminished. The stress will also be oblique to the section, as its direction must be that of the force. When the unit stress is not normal to the plane of section on which it acts, it can be decomposed into a normal unit tension and a unit shear. See § 151.

When the stress varies in magnitude from point to point, its amount on any very small area (the infinitesimal area of the Calculus) may be divided by that area, and the quotient will be the unit stress, or the amount which would exist on a square inch, if a square inch had the same stress all over it as the very small area has.

**8. Unit Stresses on Different Planes not to be Treated as Forces.**—It will be seen, upon inspection of the results of analyses which come later, that unit stresses acting on different planes must not be-compounded and resolved as if they were forces. But the entire stress upon a certain area, found by multiplying the unit stress by that area, is a force, and this force may be compounded with other forces or resolved, and the new force may then be divided by the new area of action, and a new unit stress be thus found.

Some persons may be assisted in understanding the analysis of problems by representing in a sketch, or mentally, the unit stresses at different parts of a cross-section by ordinates which make up, in their assemblage, a volume. This volume, whose base is the cross-section, will represent or be proportional to the total force on the section. The position of the resultant force or forces, *i.e.*, traversing the centre of gravity of the volume, the direction and law of distribution of the stress are then quite apparent.

# CHAPTER I.

## ACTION OF A PIECE UNDER DIRECT FORCE.

**9. Change of Length under an Applied Force.**—Let a uniform bar of steel have a moderate amount of tension applied to its two ends. It will be found, upon measurement, to have increased in length uniformly throughout the measured distance. Upon release of the tension the stretch disappears, the bar resuming its original length. A second application of the same amount of tension will cause the same elongation, and its removal will be followed by the same contraction to the original length. The bar acts like a spring. This elastic elongation (or shortening under compression) is manifested by all substances which have definite form and are used in construction; and it is the cause of such changes of shape as structures, commonly considered rigid, experience under changing loads. The product of the elongation (or shortening) into the mean force that produced it is a measure of the work done in causing the change of length. As the energy of a moving body can be overcome only by work done, the above product becomes of practical interest in structures where moving loads, shocks, and vibrations play an important part.

**10. Modulus of Elasticity.**—If the bar of steel is stretched with a greater force, but still a moderate one, it is found by careful measurement that the elongation has increased with the force; and the relationship may be laid down that the *elongation* per linear inch *is directly proportional to the unit stress* on the cross-section per square inch.

The ratio of the unit stress to the elongation per unit of length is denoted by $E$, which is termed the *modulus of elasticity* of the

## ACTION OF A PIECE UNDER DIRECT FORCE. 7

material, and is based, in English and American books, upon the pound and inch as units. If $P$ is the total tension in pounds applied to the cross-section, $S$, measured in square inches, $\lambda l$ the elongation in inches, produced by the tension, in the previously measured length of $l$ inches, and $f$ the stress per square inch of cross-section,

$$E = \frac{P}{S\lambda} = \frac{f}{\lambda}; \qquad \lambda = \frac{f}{E}.$$

Hence, if $E$ has been determined for a given material, the stretch of a given bar under a given unit stress is easily found.

Since the elongation per unit of length, $\lambda$, is merely a ratio and is the same whatever system of units is employed, $E$ will be expressed in the same units as $f$.

*Example.*—A bar of 6 sq. in. section stretches 0.085 in. in a measured length of 120 in. under a pull of 120,000 lb.

$$E = \frac{120,000 \times 120}{6 \times 0.085} = 28,200,000 \text{ lb. per sq. in.}$$

If the stress were compressive, a similar modulus would result, which will be shown presently to agree with the one just derived.

If one particle is displaced laterally with regard to its neighbor, under the action of a shearing stress, a modulus of shearing elasticity will be obtained, denoted by $C$, the ratio of the unit shear to the angle of distortion. See § 173.

**11. Stress-stretch Diagram.**—The elongations caused in a certain bar, or the stretch per unit of length, may be plotted as abscissas, and the corresponding forces producing the stretch, or the unit stresses per square inch, may be used as ordinates, defining a certain curve, as represented in Fig. 1. This curve can be drawn on paper by the specimen itself, when in the testing-machine, if the paper is moved in one direction to correspond with the movement of the poise on the weighing arm, and the pencil is moved at right angles by the stretch of the specimen.

A similar diagram can be made for a compression specimen, and may be drawn in the diagonally opposite quadrant. Pull will then be rightly represented as of opposite sign to thrust, and extension will be laid off in the opposite direction to shortening or compression.

**12. Work of Elongation.**—If the different unit stresses applied to the bar are laid off on O Y as ordinates and the resulting stretches per unit of length on O X as abscissas, the portion of the diagram near the origin will be found to be a straight line, more or less oblique, according to the scale by which the elongations are platted. The elongation varies directly as the unit stress, beginning with zero. Hence the mean force is $\frac{1}{2}P$, and the work done in stretching a given bar with a given force, if the limit of elastic stretch is not exceeded, is

$$\text{Work} = \frac{P}{2} \cdot \lambda l = \frac{P^2 l}{2ES}.$$

It may be seen that the work done in stretching the bar is represented by the area included between the base line or axis, the curve O A, and the ordinate at A. It also appears that $E$ may be looked upon as the tangent of the angle X O A. A material of greater resistance to elongation will give an angle greater than X O A and *vice versa*.

*Example.*—A bar 20 ft.=240 in. long and 3 sq. in. in section is to have a stress applied of 10,000 lb. per sq. in.; if $E=28,000,000$, the work done on the bar will be

$$\frac{30,000 \cdot 30,000 \cdot 240}{2 \cdot 28,000,000 \cdot 3} = 1,286 \text{ in.-lb.,}$$

and the stretch will be $1,286 \div 15,000 = 0.086$ in.

**13. Permanent Set.**—While the unit stress may be gradually increased with corresponding increase of stretch, and apparently complete recovery of original length when the bar is released, there comes a time when very minute and delicate measurements show that the elongation has increased in a slightly greater degree than has the stress. The line O A at and beyond such a point must therefore be a curve, concave to the axis of X. If the piece is now relieved from stress, it will be found that the bar has become permanently lengthened. The amount of this increase of length after removal of stress is called set, or *permanent set*, and the unit stress for which a permanent set can first be detected is known as the *elastic limit*. As the elongation itself is an exceedingly small quantity, even when measured in a length of many inches, and the permanent set is, in the beginning, a quantity far smaller and hence more difficult of determination, the place where the straight line O A first begins to

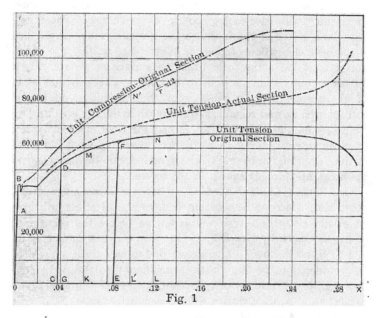

Fig. 1

curve is naturally hard to locate, and the accurate elastic limit is therefore uncertain. Some contend that O A itself is a curve of extreme flatness. The common or commercial elastic limit lies much farther up the curve, where the permanent set becomes decidedly notable.

If, after a certain amount of permanent set has occurred in a bar, and the force which caused it has been removed, a somewhat smaller force is repeatedly applied to the bar, the piece will elongate and contract elastically to the new length,

10                    STRUCTURAL MECHANICS.

*i.e.*, old length plus permanent set, just as if the unit stress were below the elastic limit.

**14. Yield-point.**—The unit stress increasing, the elongation increases and the permanent set increases until a unit stress B is reached, known as the *yield-point* (or *commercial elastic limit*, or common elastic limit), which causes the bar to yield or draw out without increase of force, and, as the section must decrease, apparently with decreasing power of resistance. There will then be a break of continuity in the graphic curve. A decided permanent elongation of the bar takes place at this time—sufficient to dislodge the scale from the surface of a steel bar, if left as it comes from the rolls or hammer. The weighing beam of the testing-machine falls, from the diminished resistance just referred to, and remains stationary while the bar is elongating for a sensible interval of time. Hence, for steel, the yield-point, or common elastic limit, is easily determined by what is known as the "drop of the beam." The remainder of the curve, up to the breaking-point, is shown in the figure.

**15. Elastic Limit Raised.**—For stresses above the yield-point also, a second application and release of stress will give an elastic elongation and contraction as before the occurrence of set, as shown by line E F, so that a new elastic limit may be said to be established. The stretch due to any given stress may be considered to be the elastic elongation plus the permanent set; and, for repetitions of lesser forces, the bar will give a line parallel to O A, as if drawn from a new origin on O X, distant from O the amount of the permanent set.

If the line O A is prolonged upwards, it will divide each abscissa into two parts, of which that on the left of O A will be the elastic stretch, and that on the right of O A the permanent set for a given unit stress.

**16. Work of Elongation, for Stress above Yield-point.**—The area below the curve, and limited by any ordinate G D, will be the work done in stretching the bar with a force represented by the product of that ordinate into the bar's cross-section, and if a line be drawn from the upper end of that ordinate parallel to O A, the triangle C D G will give the work done in elastic stretch and the quasi-parallelogram O B D C will show the permanent work of deformation done on the bar. It should be remembered that, as the bar stretches, the section decreases, and that the unit stress cannot therefore be strictly represented by $P \div S$, if $S$ is the original cross-section. The error is not of practical consequence for this discussion.

**17. Ultimate or Breaking Strength.**—If the force applied in tension to the bar is increased, a point will next be reached where a repeated application of the same *force* causes a successive increase in the permanent elongation. As this phenomenon means a gradual drawing out, final failure by pulling asunder is only a matter of a greater or less number of applications of the force. While the bar is apparently breaking under this force, the rapid diminution of cross-section near the breaking point actually gives a constantly rising unit stress, as is seen by the dotted curve of the figure.

If, however, the force is increased without pause from the beginning, the breaking force will be higher, as might be expected; since much work of deformation is done upon the bar before fracture. The bar would have broken under a somewhat smaller force, applied statically for a considerable time.

The elongation of the bar was uniform per unit of length during the earlier part of the test. There comes a time when a portion or section of the bar, from some local cause, begins to yield more rapidly than the rest. At once the unit stress at that section becomes greater than in the rest of the bar, by reason of decrease of cross-section, and the drawing out becomes intensified, with the result of a great local elongation and necking of the specimen and an assured final fracture at that place. If the bar were perfectly homogeneous, and the stress uniformly distributed, the bar ought to break at the middle of the length, where the *flow* of the metal is most free.

It is customary to determine, and to require by specification, in addition to elastic limit and ultimate strength (on one continuous application of increasing load), the per cent. of elongation after fracture (which is strictly the permanent set) in a certain original measured length, usually eight inches, and the per cent. of reduction of the original area, after fracture, at the point of fracture. As the measured length must include the much contracted neck, the *average* per cent. of elongation is given under these conditions. A few inches excluding the neck would show less extension, and an inch or two at the neck would give a far higher per cent. of elongation. The area between the axis of $X$, the extreme ordinate, and the curve will be the work of fracture, if $S$ is considered constant, and will be a measure of the ability of the material to resist shocks, blows, and vibrations before fracture. It is indicative of the toughness or ductility of the material.

The actual curve described by the autographic attachment to a testing-machine is represented by the full line; the real relation of stress per square inch to the elongation produced, when account is taken of the progressive reduction of sectional area, is shown by the dotted line. The yield-point, or common elastic limit, is very marked, there appearing to be a decided giving way or rearrangement of the particles at that value of stress. The true elastic limit is much below that point.

**18. Effect of a Varying Cross-section.**—If a test specimen is reduced to a smaller cross-section, by cutting out a curved surface, for only a short distance as compared with its transverse dimensions, it will show a greater unit breaking stress, as the metal does not flow freely, and lateral contraction of area is hindered. But, if the portion of reduced cross-section joins the rest of the bar by a shoulder, the apparent strength is reduced, owing to a concentration of stress on the particles at the corner as the unit stress suddenly changes from the smaller value on the larger section to the greater unit stress on the smaller cross-section.

**19. Compression Curve.**—A piece subjected to compression will shorten, the particles being forced nearer together, and the cross-section will increase. It might be expected, and is found by experiment to be the case, that, in the beginning, the resistance of the particles to approach would be like their resistance to separation under tension, so that the tension diagram might be prolonged through the origin into the third quadrant, reversing

## ACTION OF A PIECE UNDER DIRECT FORCE. 13

the sign of the ordinate which represents unit stress and of the abscissa which shows the corresponding change of length. As this part of the diagram is a straight line, it follows that the value of $E$, the elastic modulus for compression, is the same as that for tension. After passing the yield-point the phenomena of compression are not so readily determined, as fracture or failure by compressive stress is not a simple matter, and the increase of sectional area in a short column of ductile material will interfere with the experiment. In long columns and with materials not ductile, failure takes place in other ways, as will be explained later.

The compression curve is here shown in the same quadrant with the tension curve for convenience and comparison.

**20. Resilience.**—By definition, § 10, if $f$ is the unit stress per square inch and $\lambda l$ the stretch of a bar of length $l$, in inches, the modulus of elasticity $E = f \div \lambda$, provided $f$ does not exceed the elastic limit. Also the work done in stretching a bar inside the elastic limit, by a force $P$, gradually applied, that is, beginning with zero and increasing with the stretch, is the product of the mean force, $\tfrac{1}{2} P$, into the stretch, or

$$\text{Work done} = \tfrac{1}{2} P \cdot \lambda l = \frac{f S}{2} \cdot \frac{fl}{E} = \frac{f^2}{2E} \cdot Sl.$$

The amount of work which must be done upon a piece in order to produce the safe unit stress, $f$, in it is the *resilience* of the piece. $Sl$ is the volume of the bar; $f^2 \div E$ is called the modulus of resilience, when $f$ is the elastic limit, or sometimes the maximum safe unit stress. This modulus depends upon the quality of the material, and, as it is directly proportional to the amount of work that can safely be done upon the bar by a load, it is a measure of the capacity of a certain material for resisting or absorbing shock and impact without damage. For a particular piece, the volume $Sl$ is also a factor as above. A light structure will suffer more from sudden or rapid loading than will a heavier one of the same material, if proportioned for the same unit stress.

**21. Work Done Beyond the Elastic Limit.**—The work done in stretching a bar to any extent is, in Fig. 1, the area in the diagram between the curve from the origin up to any point, the ordinate to that point, and the axis of abscissas, provided the ordinate represents $P$, and the abscissa the total stretch.

Further, it may be seen from the figure that, if a load applied to the bar has exceeded the yield-point, the bar, in afterwards contracting, follows the line F E; and, upon a second application of the load, the right triangle of which this line is the hypotenuse will be the work done in the second application, a smaller quantity than for the first application. But, if the load, in its second and subsequent applications, possesses a certain amount of energy, by reason of not being gently or slowly applied, this energy may exceed the area of the triangle last referred to, with the result that the stress on the particles of the bar may become greater than on the first application. Indeed it is conceivable that this load may be applied in such a way that the resulting unit stress may mount higher and higher with repeated applications of load, until the bar is broken with an apparent unit stress $P \div S$, far less than the ultimate strength, and one which at first was not much above the yield-point. If the load in its first application is above the yield-point of the material, and it is repeated continuously, rupture will finally occur.

What is true for tensile stresses is equally true for compressive stresses, except that the ultimate strength of ductile materials under compression is uncertain and rather indefinite.

**22. Sudden Application of Load.**—If a steel rod, 10 feet = 120 inches long, and one square inch in section, with $E = 28{,}000{,}000$, is subjected to a force increasing gradually from 0 to 12,000 lb. longitudinal tension, its stretch will be $12{,}000 \times 120 \div 28{,}000{,}000 = 0.05$ in., and the work done in stretching the bar will be $\frac{1}{2} \times 12{,}000 \times 0.05 = 300$ in.-lb.

But if the 12,000 lb. is suddenly applied, as by the extremely rapid loading of a structure of which the rod forms a part, or by the quick removal of a support which held this weight at the lower end of the rod, the energy due to a fall of 0.05 in. is $12{,}000 \times 0.05 = 600$ in.-lb., while the work done upon the rod is but 300 in.-lb.

as before. The excess of 300 in.-lb. of energy must be absorbed by the rod and it will continue to stretch until the energy due to the fall equals the work done upon the rod or until it has stretched 0.10 in. The stress in the rod is then 24,000 lb. or twice the suddenly applied load; the energy due to the fall is $12,000 \times 0.10$ and the work done upon the rod is $\frac{1}{2} \times 24,000 \times 0.10 = 1,200$ in.-lb. As equilibrium does not exist between the external force and the internal stress, the rod will contract and then undergo a series of longitudinal vibrations of decreasing amplitudes, finally settling down to a stretch of 0.05 in., when the extra work of acceleration has been absorbed. The work of acceleration on the mass of the bar is neglected.

A load applied to a piece with absolute suddenness produces twice the deformation and twice the stress which the same load does if applied gradually. Stresses produced by moving loads on a structure are intermediate in effect between these two extremes, depending upon rapidity or suddenness of loading. Hence it is seen why the practice has arisen of limiting stresses due to moving loads *apparently* to only one-half of the values permitted for those caused by static loading.

For resilience or work done in deflection of beams, see § 100.

**23. Granular Substances under Compression.**—Failure by Shearing on Oblique Planes. Blocks of material, such as cast iron, sandstone, or concrete, when subjected to compression, frequently give way by fracturing on one or more oblique planes which cut the block into two wedges, or into pyramids and wedges. The pyramids may overlap, and their bases are in the upper and lower faces of the block. This mode of fracture, peculiar to granular substances, of comparatively low shearing resistance, can be discussed as follows:

If a short column, Fig. 2, of cross-section $S$ is loaded centrally with $P$, the unit compression on the right section will be $p_1 = P \div S$, and if the short column gives way under this load, this value of $p_1$ is commonly considered the crushing strength of the material. While it doubtless is the *available* crushing strength of this specimen, it may by no means represent the maximum resistance to crushing under other conditions.

If $p_1 = P \div S$ is the unit thrust on the right section, it is seen, from § 151, that, on a plane making an angle $\theta$ with the right section, the normal unit stress $p_n = p_1 \cos^2 \theta$, and the tangential unit stress $q = p_1 \sin \theta \cos \theta$. If $m$ = coefficient of frictional resistance of the material to sliding, the resistance per square inch

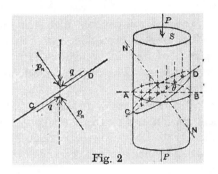

Fig. 2

to sliding along this oblique plane will be $mp_n = mp_1 \cos^2 \theta$, and the portion of the unit shearing stress tending to produce fracture along this plane will be $q - mp_n = p_1 (\sin \theta \cos \theta - m \cos^2 \theta)$.

Fracture by shearing, if it occurs, will take place along that plane for which the above expression is a maximum, or $d(q - mp_n) \div d\theta = 0$. Differentiating relatively to $\theta$,

$$p_1(\cos^2 \theta - \sin^2 \theta + 2m \sin \theta \cos \theta) = 0;$$

$$\sin^2 \theta - 2m \sin \theta \cos \theta + m^2 \cos^2 \theta = \cos^2 \theta + m^2 \cos^2 \theta;$$

$$\sin \theta - m \cos \theta = \cos \theta \sqrt{(1 + m^2)};$$

$$\frac{\sin \theta}{\cos \theta} = \tan \theta = m + \sqrt{(1 + m^2)}.$$

If $m$ were zero, $\theta$ max. would be 45°. Therefore the plane of fracture always makes an angle greater than 45° with the right section. As $\theta$ may be negative as well as positive, fracture tends to form pyramids or cones.

*Example.*—A rectangular prism of cast iron, 2 in. high and square section = 1.05 sq. in., sheared off under a load of 97,000 lb., or 92,380

## ACTION OF A PIECE UNDER DIRECT FORCE. 17

lb. compression per sq. in. of cross-section, at an angle whose tangent was 1.5, or 56° 19′.

$$1.5 = m + \sqrt{(1+m^2)}; \quad 2.25 - 3m + m^2 = 1 + m^2; \quad m = 0.42.$$
$$\sin \theta = 0.8321, \quad \cos \theta = 0.5546, \quad \sin \theta \cos \theta = 0.461, \quad \cos^2 \theta = 0.308.$$
$$92{,}380(0.461 - 0.42 \times 0.308) = 30{,}680 \text{ lb.}$$

The coefficient of friction is 0.42, and the shear 30,680 lb. per sq. in. The crushing strength of a short block would have exceeded considerably the above 92,400 lb.

Since this deviation of the plane of fracture from 45° is due to a resistance analogous to friction, it follows that, when a column of granular material, and of moderate length, gives way by shearing, the value $p_1$ will be only that compressive stress which is compatible with the unit shearing strength, while its real compressive strength in large blocks will be much higher.

The same phenomenon is exhibited by blocks of sandstone and of concrete. Tests of cubes and flat pieces yield higher results than do those of prisms of the same cross-section and having a considerably greater height.

**24. Ductile Substances under Compression.**—Wrought iron, and soft and medium steel, as well as other ductile substances, tested in short blocks in compression, bulge or swell in transverse dimensions, and do not fracture. Hence the ultimate compressive strength is indefinite.

**25. Fibrous Substances under Compression.**— Wood and fibrous substances which have but small lateral cohesion of the fibres, when compressed in short pieces in the direction of the same, separate into component fibres at some irregular section, and the several fibres fail laterally and crush.

**26. Vitreous Substances under Compression.**—Vitreous substances, like glass and vitrified bricks, tend to split in the direction of the applied force.

**27. Resistance of Large Blocks.**—The resistance per square inch of a cube to compression will depend upon the size of the cube. As the unit stress and the resulting deformation are associated, as noted in § 4, it follows that the unit compressive stress will be greatest at the centre of the compressed surface and least at the free edges where lateral movement of the particles is less restrained. Hence, the larger the cube, the greater the mean or apparent strength per square inch. Large blocks of stone, therefore, have a greater average sustaining power per square inch than is indicated by small test specimens, other things being equal.

The same inference can be drawn as to resistance of short pieces to tension as compared with longer pieces of the same cross-section.

A uniform compression over any cross-section of a large post or masonry pier, when the load is centrally applied to but a small portion of the top can be realized only approximately; the same thing is probably true of the foundation below the pier. The resisting capacity of the material, if earth, is thereby enhanced; for the tendency to escape laterally at the edges of the foundation is not so great as would be the case if the load were equally severe over the whole base.

Beveling the edges of the compressed face of a block will increase the apparent resistance of the material by taking the load from the part least able to stand the pressure. The unloaded perimeter may then act like a hoop to the remainder.

*Examples.*—1. A round bar, 1 in. in diameter and 10 ft. long, stretches 0.06 in., under a pull of 10,000 lb. What is the value of $E$? What is the work done?  25,464,733; 300 in.-lb.

2. If the elastic limit of the bar is reached by a tension of 30,000 lb. per sq. in., what is the work done or the resilience of the bar?
1,666 in.-lb.

3. An iron rod, $E=29,000,000$, hangs in a shaft 1,500 ft. deep. What will be the stretch?  1.55 in.

4. A certain rod, 22 ft. long, and having $E=28,000,000$, is to be adjusted by a nut of 8 threads to the inch to an initial tension of 10,000 lb. per sq. in. If the connections were rigid, how much of a revolution ought to be given to the nut after it fairly bears?  0.75.

5. Can a weight of 20,000 lb. be lifted by cooling a steel bar, 1 in. sq., from 212° to 62° F.? Coefficient of expansion=0.0012 for 180°; $E=29,000,000$.

6. A steel eye-bar, 80 in. long and 2 in. sq., fits on a pin at each end with $\frac{1}{50}$ in. play. What will be the tension in the bar, if the temperature falls 75° F. and the pins do not yield?
7,250 lb. per sq. in.

7. A cross-grained stick of pine, 1 sq. in. in section, sheared off at an angle of about 66° with the right section under a compressive load of 3,200 lb. If the coefficient of friction is 0.5, what is the unit shearing stress of the section, the actual irregular area being 2.9 sq. in.?
780 lb.

## CHAPTER II.

### MATERIALS.

**28. Growth of Trees.**—Trees from which lumber is cut grow by the formation of woody fibre between the trunk and the bark, and each annual addition is more or less distinctly visible as a ring. Each ring is made up of a light, porous part, the spring wood, and a darker, dense part, the summer wood. Since the latter is firm and heavy it determines to a large extent the weight and strength of the timber. The sap circulates through the newer wood, and in most trees the heart-wood, as it is called, can be easily distinguished from the sap-wood. The former is considered more strong and durable, unless the tree has passed its prime. The heart then deteriorates. Sap-wood, in timber exposed to the weather, is the first to decay.

Branches increase in size by the addition of rings, as does the trunk; hence a knot is formed at the junction of the branch with the trunk. The knot begins where the original bud started, and increases in diameter towards the exterior of the trunk, as the branch grows. The grain of the annual growth, formed around the junction of the branch with the trunk, is much distorted. Hence timber that contains large knots is very much weaker than straight-grained timber. Even small knots determine the point of fracture when timber is experimentally tested for strength. When a branch happens to die, but the stub remains, and annual rings are added to the trunk, a dead or loose knot occurs in the sawed timber; such a knot is considered a defect, as likely to let in moisture and start decay.

As forest trees grow close together, the branches die successively from below from lack of sunlight; such trees develop straight trunks of but little taper, free from any knots, except insignificant ones immediately around the centre, and yield straight-grained, clear lumber. A few trees, like hemlock, sometimes have their fibres running in a spiral, and hence yield cross-grained timber. Trees that grow in open spaces have large side limbs, and the lumber cut from them has large knots.

**29. Shrinkage of Timber.**—If a log is stripped of its bark and allowed to dry or season, it will be found that the contraction or shrinkage in the direction of the radius is practically nothing. There are numerous bundles or ribbons of hard tissue running radially through the annual rings which appear to prevent such shrinkage. Radial cracks, running in to a greater or less distance, indicate that the several rings have yielded to the tension set up by the tendency to shrink circumferentially. Sawed timber of any size is likely to exhibit these season cracks. Such cracks are blemishes and may weaken the timber when used for columns or beams. By slow drying, and by boring a hole through the axis to promote drying within, the tendency to form season cracks may be diminished.

A board sawed radially from a log will not shrink in width, and will resist wear in a floor. Such lumber is known as *quarter-sawed*. A board taken off near the slab will shrink much and will tend to warp or become concave on that side which faced the exterior of the log. For that reason, and because the annual rings have less adhesion than the individual fibres have, all boards exposed to wear, as in floors, should be laid heart-side down.

**30. Decay of Wood.**—Timber exposed to the weather should be so framed together, if possible, that water will not collect in joints and mortises, and that air may have ready access to all parts, to promote rapid drying after rain. The end of the grain should not be exposed to the direct entrance of water, but should be covered, or so sloped that water can run off, and the ends should be stopped with paint. It is well to paint joints before they are put together.

The decay of timber is due to the presence and action of vegetable growths or fungi, the spores of which find lodgment in the pores of the wood, but require air and moisture, with a suitable temperature, for their germination and spread. Hence if timber is kept perfectly dry it will last indefinitely. If it is entirely immersed in water, it will also endure, as air is excluded. Moisture may be excluded from an exposed surface by the use of paint. Unseasoned timber painted, or placed where there is no circulation of air, will dry-rot rapidly in the interior of the stick; but the exterior shell will be preserved, since it dries out or seasons to a little depth very soon.

The worst location for timber is at or near the ground surface; it is then continually damp and rot spreads fast.

**31. Preservation of Wood.**—The artificial treatment of timber to guard against decay may be briefly described as the introduction into the pores of some poison or antiseptic to prevent the germination of the spores; such treatment is efficacious as long as the substance introduced remains in the wood. Creosote is the best of preservatives and the only one effective against sea-worms, but is expensive.

The timber is placed in a closed tank and steam is admitted to soften the cells. After some time the steam is shut off, a partial vacuum is formed, and the preservative fluid is run in and pressure applied to force the liquid into the pores of the wood. As steaming injures the fibres, treated timber is weaker than untreated.

Burnettizing is the name given to treatment with zinc chloride, a comparatively cheap process, applied to railway-ties and paving-blocks. To prevent the zinc chloride from dissolving out in wet situations, tannin has been added after the zinc, to form with the vegetable albumen a sort of artificial leather, plugging up the pores; hence the name, zinc-tannin process. For bridge-timbers burnettizing makes the timber unduly brittle.

As the outside of treated timber contains most of the preservative, timber should be framed before being treated.

**32. Strength of Timber.**—The properties and strength of different pieces of timber of the same species are very variable. Seasoned lumber is much stronger than green, and of two pieces of the same species and of the same dryness, the heavier is the stronger, while, in general, heavy woods are stronger than light.

Prudence would dictate that structures should be designed for the strength of green or moderately seasoned timber of average quality. As the common woods have a comparatively low resistance to compression across the grain, particular attention should be paid to providing sufficient bearing area where a strut or post abuts on the side of another timber. An indentation of the wood destroys the fibre and increases the liability to decay, if the timber is exposed to the weather, especially under the continued working produced by moving loads.

Average breaking stresses of some American timbers as found by the Forestry Div. of the U. S. Dept. of Agriculture follow. The results are for well-seasoned lumber, 12 per cent. moisture:

|  | Weight per Cu. Ft. | Compression. | | Bending. | | Shear with Grain. |
|---|---|---|---|---|---|---|
|  |  | With Grain. | Across Grain. | Modulus of Rupture. | Modulus of Elasticity. |  |
| Long-leaf pine..... | 38 | 8,000 | 1,260 | 12,600 | 2,070,000 | 835 |
| Short-leaf pine..... | 32 | 6,500 | 1,050 | 10,100 | 1,680,000 | 770 |
| White pine........ | 24 | 5,400 | 700 | 7,900 | 1,390,000 | 400 |
| Douglas spruce.... | 32 | 5,700 | 800 | 7,900 | 1,680,000 | 500 |
| White oak......... | 50 | 8,500 | 2,200 | 13,100 | 2,090,000 | 1,000 |

See also § 145.

Timber is graded or classified at the sawmills according to the standard rules of different manufacturers' associations, and specifications should call for a grade of lumber agreeing with the classification of the mills of the region where the lumber is produced.

**33. Iron and Carbon.**—Cast iron and steel differ from each other in physical qualities on account of the different percentages of carbon in combination with the iron. Ordinary cast or pig iron contains from $3\frac{1}{2}$ to 4 per cent. of carbon, while structural steel contains from one to two tenths of one per cent. Wrought iron and steel are made by removing the metalloids from cast iron.

**34. Cast Iron.**—Iron ore, which is an oxide of iron, is put in a blast-furnace together with limestone and coke. Superheated air is blown in at the bottom of the furnace and the burning of the coke produces a high temperature and removes the oxygen from the ore. The earthy materials in the ore unite with the limestone and form a slag which floats on the surface of the molten metal and is drawn off separately. The iron is run off into molds and forms pig iron.

When broken, the pig is seen to be crystalline and its color may be white or gray, depending upon the condition of the carbon in the iron. In the furnace the carbon is dissolved in the bath, but when the iron solidifies the carbon may either remain in solution and produce *white* iron, or part of the carbon may precipitate in the form of scales of graphite and produce *gray* iron. The condition of the carbon depends partly on the rate of cooling, but more on the other elements present. White iron is hard and brittle; gray iron is tougher. If gray iron is run into a mold lined with iron, it is *chilled* from the surface to a depth of one-half to three-fourths of an inch; that is, the surface is turned to white iron and made intensely hard, as in the treads of car-wheels.

Besides the carbon, pig iron contains more or less silicon, usually from one to two per cent. It tends to make the carbon take the graphitic form. Sulphur makes the iron hard and brittle; good foundry iron should not contain more than 0.15 of one per cent. Phosphorus makes molten iron fluid, and irons high in phosphorus are used in making thin and intricate castings, but such castings are very brittle. The amount of silicon and sulphur in pig iron can be controlled by the furnaceman, but the only way in which the amount of phosphorus can be kept down is by using pure ores and fuel. The phosphorus in pig iron to be used for making steel by the acid Bessemer process is limited to one-tenth of one per cent.; iron fulfilling that requirement is called Bessemer pig.

The tensile strength of cast iron varies from 15,000 to 35,000 lb. and the compressive strength from 60,000 to 200,000 lb. per sq. in. The modulus of elasticity ranges from 10 to 30 million pounds per square inch. For ordinary foundry iron the tensile strength is usually 18,000 to 22,000 lb. and the modulus of elasticity 12 to 15 million. As cast iron is brittle and likely to contain hidden defects, it is little used in structural work.

**35. Wrought Iron.**—Wrought iron is made by melting pig iron, and cinder which contains oxide of iron, together in a reverberatory furnace. The carbon and silicon in the iron unite with the oxygen of the slag, leaving metallic iron. As the carbon is removed the melting-point rises, and since the temperature of the furnace is not high enough to keep the iron fluid, it assumes a pasty condition. The semi-fluid iron is collected into a lump by the puddler and withdrawn from the furnace. It is then much like a sponge; the particles of wrought iron have adhered to one another, but each particle of iron is more or less coated with a thin film of slag and oxide, as water is spread through the pores of a partly dry sponge.

The lump of iron is put into a squeezer, and the fluid slag and oxide drip out as water does from a squeezed sponge. But, as it is impracticable to squeeze a sponge perfectly dry, so it is impracticable to squeeze all the impurities out from among the particles of metallic iron. In the subsequent processes of rolling and re-rolling each globule of iron is elongated, but the slag and oxide are still there; so that the rolled bar consists of a collection of threads of iron, the adhesion of which to each other is not so great as the strength of the threads.

If the surface of an iron bar is planed smooth and then etched with acid, the metal is dissolved from the surface and the black lines of impurities are left distinctly visible.

That wrought iron is fibrous is then an accident of the process of manufacture, and does not add to its strength. If these impurities had not been in the iron when it was rolled out, it would have been more homogeneous and stronger. The fibrous fracture of a bar which is nicked on one side and broken by bending is not especially indicative of toughness; for soft steel is tough and ductile without being fibrous.

The tensile strength of wrought iron is about 50,000 lb. per sq. in. and its modulus of elasticity about 28,000,000. Wrought iron is still used to some extent when it is necessary to weld the material, but soft steel has largely driven it from the market.

**36. Steel.**—Structural steel is made from pig iron by burning out the metalloids.

In the *Bessemer* process molten pig iron is run into a converter and cold air is blown through the metal to burn out the carbon and silicon. The combustion generates enough heat to keep the metal fluid, although the melting-point rises. When the metal is free from carbon and silicon, manganese is added to remove the oxygen dissolved in the metal and to make the steel tough when hot. If it is desired to add carbon to make a stronger steel, spiegel-iron is used, which is a pig iron containing about 12 per cent. manganese and 4 or 5 per cent. carbon. After the manganese is added the metal is cast into an ingot.

In America the *acid* Bessemer process is used exclusively; that is, the converters are lined with a silicious material which necessitates a silicious or acid slag, and consequently there is no elimination of sulphur and phosphorus, which can enter a basic slag only. This process is the cheapest way of making steel, but the product is not so uniform as that of the open-hearth furnace. Bessemer steel is used for rails and for buildings, but is not used for first-class structural work.

An *open-hearth* furnace is a regenerative furnace having a hearth for the metal exposed to the flame. Heat is generated by burning gas which, together with the air-supply, has been heated before it is admitted to the furnace. An intense heat is thus produced which is sufficient to keep the metal fluid after the carbon has been removed. The furnace is charged with pig iron and scrap and, after the charge is melted, iron ore is added. The carbon and silicon unite with the oxygen of the air and of the ore. When the required composition is attained the steel is drawn off, ferromanganese is thrown into the ladle and the ingot is cast. After removal from the molds the ingots are heated and rolled into plates or shapes. Steel castings are made by pouring the metal from the furnace or the converter into molds.

If the hearth is lined with sand, the slag formed during the oxidation is silicious or *acid* and the oxide of phosphorus, which acts as an acid, reunites with the iron. If, however, the hearth is lined with dolomite, limestone may be added to the charge to form a *basic* slag which the phosphorus may enter. The sulphur also may be reduced to some extent by the basic process, but not by the acid.

The tensile strength of pure iron is probably about 40,000 lb. per sq. in., but the presence of other elements, as carbon and phosphorus in small quantities, increases its strength and makes it more brittle. The element which increases the strength most with the least sacrifice of toughness is carbon, and it is the element which the manufacturer uses to give strength. In structural steels it may range from 0.05 to 0.25 of one per cent. Phosphorus and sulphur are kept as low as possible. Phosphorus makes the steel brittle at ordinary temperatures, while sulphur makes it brittle at high temperatures and likely to crack when rolled. Manganese makes the steel tough while hot. It ranges from 0.30 to 0.60 of one per cent. in ordinary structural steels.

The strength of rolled steel depends somewhat upon the thickness of the material, thin plates which have had more work done upon them being stronger. The softer structural steels can be welded readily, and the medium with care. They will not temper. The modulus of elasticity of both soft and medium steel is about 29,000,000 lb. per sq. in.

**37. Classification of Steels.**—The following classification and requirements are taken from the standard specifications of the American Society for Testing Materials.

| Steel. | Tensile Strength in 1000 Lb. | Elongation in 8 In. not Less than | P Acid | P Basic | S | Mn |
|---|---|---|---|---|---|---|
| | | | \multicolumn{3}{c|}{not more than} | |

| | | | | | | |
|---|---|---|---|---|---|---|
| \multicolumn{7}{c}{STRUCTURAL STEEL FOR BRIDGES AND SHIPS.} | 
| Rivet........ | 50 to 60 | 26% | 0.08% | 0.06% | 0.06% | |
| Soft.......... | 52 to 62 | 25% | 0.08% | 0.06% | 0.06% | |
| Medium...... | 60 to 70 | 22% | 0.08% | 0.06% | 0.06% | |
| \multicolumn{7}{c}{STRUCTURAL STEEL FOR BUILDINGS.} |
| Rivet........ | 50 to 60 | 26% | 0.10% | 0.10% | | |
| Medium...... | 60 to 70 | 22% | 0.10% | 0.10% | | |

## MATERIALS.

**BOILER-PLATE AND RIVET-STEEL.**

| Steel. | Tensile Strength in 1000 Lb. | Elongation in 8 In. not Less than | P Acid | P Basic | S | Mn |
|---|---|---|---|---|---|---|
| | | | not more than | | | |
| Extra-soft | 45 to 55 | 28% | 0.04% | 0.04% | 0.04% | 0.30 to 0.50% |
| Fire-box | 52 to 62 | 26% | 0.04% | 0.03% | 0.04% | 0.30 to 0.50% |
| Boiler | 55 to 65 | 25% | 0.06% | 0.04% | 0.05% | 0.30 to 0.60% |

Steel for buildings may be made by the Bessemer process; the other two classes must be made by the open-hearth process. Test specimens of rivet and soft steel not more than three-fourths of an inch thick must bend cold 180° flat without fracture; similar specimens of medium steel must bend cold 180° around a diameter equal to the thickness of the piece without fracture. The yield-point must not be less than one-half the ultimate strength.

This classification gives a good idea of the usual requirements for steel, although some engineers prefer a grade of structural steel midway between soft and medium, that is, one having a strength of 55 to 65 thousand pounds per square inch. Such a grade is recommended by the American Railway Engineering and Maintenance-of-Way Association.

**38. Work of Elongation.**—It is seen from the diagram Fig. 1 that the resistance of the metal per square inch increases as the bar draws out, and diminishes in section under tension, as shown by the dotted curve, although the total resistance grows less near the close of the test, as shown by the full line. As a small increase in the amount of carbon diminishes the elongation and reduction of area, it is possible that the carbon affects the apparent ultimate strength in this manner (since such strength is computed on the square inch of original section), and not by actually raising the resisting power of the metal.

Since the measure of the work done in stretching a bar is the product of one-half the force by the stretch, if the yield-point has not been passed, and, for values beyond that point, is the area below the curve in the diagram, limited by the ordinate representing the maximum force, the comparative ability of a material to resist live load, shock, and vibration is indicated by this area. A mild steel of moderate strength may thus have greater value than a higher carbon steel of much greater tensile strength.

**39. Tool-steel.**—Tool-steel as well as spring-steel of good quality is made by melting wrought iron or steel of known composition in a crucible, and may contain from one-half to one per cent. of carbon. When heated a bright red and quenched in water such steel becomes very hard and brittle and entirely loses the property of drawing out; but if it is subsequently heated to a moderate temperature and then allowed to cool slowly, its strength is increased and its brittleness reduced, while it still retains more or less of its hardness. This process is called *tempering*. Springs and tools are tempered before being used. Some special tool-steels contain tungsten or chromium, which give great hardness without tempering.

**40. Malleable Iron: Case-hardened Iron.**—There are two other products which may well be mentioned, and which will be seen to unite or fit in between the three already described. The first is what is known as "malleable cast iron" or malleable iron.

Small articles, thin and of irregular shapes, which may be more readily cast than forged or fashioned by a machine, and which need not be very strong, are made of white cast iron, and then imbedded in a substance rich in oxygen, as, for instance, powdered red hematite iron ore, sealed up in an iron box, and heated to a high temperature for some time. The oxygen abstracts the carbon from the metal to a slight depth, converting the exterior into soft iron, while the carbon in the interior takes on a graphitic form with an increase of strength and diminution of brittleness.

The second product is case-hardened iron. An article fashioned of wrought iron or soft steel is buried in powdered charcoal and heated. The exterior absorbs carbon and is converted into high steel, which will better resist wear and violence than will soft iron. The Harvey process for hardening the exterior of steel armor-plates is of a similar nature.

**41. Effect of Shearing and Punching.**—As was shown in § 15, when a bar of steel is stressed beyond the elastic limit and has received a permanent set, a higher elastic limit is established, but the percentage of elongation is much reduced, as shown by the curve E F N of Fig. 1. The steel therefore has been hardened in the sense that its ductility has been lessened. Examples of this hardening are seen in plates which are rolled cold and in drawn wire. Similarly when a rivet-hole is punched in a plate, the metal immediately surrounding the hole is distorted and hardened, thus reducing the ductility of the plate around the hole. If the strip containing a punched hole is loaded beyond the elastic limit of the plate, the metal surrounding the hole, being unable to stretch as much as the rest, is unduly stressed and the ultimate strength is less than it would have been had the stress been uniformly distributed. Experiments show that plates with punched holes are weaker than those with drilled holes.

The same hardening effect is produced by shearing, or cutting, a plate. When a bar of punched or sheared steel is bent, cracks form at the hard edges and spread across the plate; but if the holes are reamed out or the sheared edges are planed off to a small depth, the hardened metal is removed and the bar will bend without cracking. Medium steel, especially if thick, is injured much more than soft steel by punching and shearing. Specifications for structural work frequently require rivet-holes to be reamed to a diameter three-sixteenths of an inch larger than the punch, and one-quarter of an inch to be planed from the edges of sheared plates. In good boiler-work the rivet-holes are drilled.

The ductility of steel which has been hardened by cold working can be restored by *annealing*, that is, by heating to a red heat and then cooling slowly.

**42. Building Stone.**—The principal building stones may be grouped as granites, limestones, and sandstones. *Granite* consists of crystals of quartz, felspar and mica or hornblende. It is very strong and durable, but its hardness makes it difficult to work. Owing to its composite structure it does not resist fire well. *Limestone* is a stratified rock of which carbonate of lime is the chief ingredient. When limestone consists of nearly pure carbonate and is of good color and texture, it is called marble. *Sandstone* consists of grains of sand cemented together by silica, carbonate of lime, iron oxide, or clay. If the cementing material is silica, the stone is very hard to work. Sandstone is one of the most valuable of building materials.

Sound, hard stones like granite, compact limestone, and the better grades of sandstone are sufficiently strong to carry any loads brought upon them in ordinary buildings; hence the question of durability rather than strength is the governing consideration in selecting a good building stone. The only sure test of the ability of a building stone to resist climatic changes, to stand the weather, is the lapse of time. Artificial freezing and thawing of a small specimen, frequently repeated, will give indication as to durability.

Stratified stones should be laid on their natural beds, that is, so that the pressure shall come practically perpendicular to the layers. They are much stronger in such a position, and the moisture which porous stones absorb from the rain can readily dry out. If the stones are set on edge, the moisture is retained and, in the winter season, tends to dislodge fragments by the expansive force exerted when it freezes. Some sandstone facings rapidly deteriorate from this cause. Crystals of iron pyrites occur in some sandstones and unfit them for use in the face of walls. The discoloration resulting from their oxidation, and the local breaking of the stone from the swelling are objectionable.

The modulus of elasticity differs greatly for different stones. Limestones and granites are nearly perfectly elastic for all working loads, but sandstones take a permanent set for the smallest loads. Tests of American building stones in compression made at the Watertown Arsenal give values of 5 to 10 million for granites and marbles and 1 to 3 million for sandstones. The weight of granite ranges from 160 to 180, of limestone and marble from 150 to 170, and of sandstone from 130 to 150 pounds per cubic foot.

**43. Masonry.**—Most masonry consists of regularly coursed stones on the face, with a backing of irregular-shaped stones behind. Stones cut to regular form and laid in courses make *ashlar* masonry, if the stones are large and the courses continuous. When the stones are smaller, and the courses not entirely continuous, or sometimes quite irregular, although the faces are still rectangular, the descriptive name is somewhat uncertain, as block-in-course, random range, etc., down to coursed rubble, where the end joints of the stones are not perpendicular to the beds. *Rubble* masonry denotes that class where the stones are of irregular shape, and fitted together without cutting. If the face of the stone is left as it comes from the quarry, the work is called quarry-faced or rock-faced. The kind of masonry depends upon the beds and joints. Walls of stone buildings have only a more or less thin facing of stone, the body of the wall being of brick. The stone facing should be well anchored to the brickwork by iron straps.

44. **Bricks.**—Bricks are made from clay which may be roughly stated to be silicate of alumina ($Al_2O_3, 2SiO_2, 2H_2O$). The clay is freed from pebbles, mixed with water in a pug-mill and molded. The green bricks are dried in the air and then burned in kilns. *Pressed* bricks are pressed after drying. They have a smooth exterior, are denser and are more expensive than common bricks. *Paving*-bricks are made from hard, laminated, rock-like clays called shales, which are not plastic unless pulverized and mixed with water. Paving-bricks are burned to incipient vitrification, which makes them extremely hard. Lime and iron in clay act as fluxes and make the clay fusible; *fire*-bricks are therefore made from clay free from fluxes. If limestone pebbles occur in a brick-clay, they must be removed or they will form lumps of lime after burning, and when wet will slake, swell, and break the bricks.

The red color of common bricks is an accidental characteristic, due to iron in the clay. Such bricks are redder the harder they are burned, finally, in some cases, turning blue. The cream-colored bricks with no iron may be just as strong and are common in some sections. Soft, underburned bricks are very porous, absorb much water, and cannot be used on the outside of a wall, especially near the ground line, for they soon disintegrate from freezing. Hard-burned bricks are very strong and satisfactory in any place; they can safely carry six or eight tons to the square foot. Bricks differ much in size in different parts of the country. A good brick should be straight and sharp-edged, reasonably homogeneous when broken, dense and heavy. Two bricks struck together should give a ringing sound.

*Sand bricks* are made by mixing thoroughly sand with 5 or 10 per cent. of slaked lime and sufficient water to allow molding. The bricks are formed under very great pressure and are then run into a large boiler and exposed to the action of steam under pressure for several hours. Some chemical reaction takes place between the silica and the lime under the conditions of heat and moisture, which firmly cements the particles of sand. Well-made sand bricks have a crushing strength of 2,500 to 5,000 lb. per sq. in. They are denser than common bricks and are very regular in shape and size.

**45. Lime.**—Lime for use in ordinary masonry and brickwork is made by burning limestone, or calcium carbonate, $CaCO_3$, and thus driving off by a high heat the carbon dioxide and such water as the stone contains. There remains the quicklime of commerce, $CaO$, in lumps and powder. This quicklime has a great affinity for water and rapidly takes it up when offered, swelling greatly and falling apart, or slaking, into a fine, dry, white powder, $Ca(OH)_2$, with an evolution of much heat, due to the combination of the lime with the water. The use of more water produces a paste, and the addition of sand, which should be silicious, sharp in grain and clean, makes lime mortar. The sand is used partly for economy, partly to diminish the tendency to crack when the mortar dries and hardens, and partly to increase the crushing strength. The proportion is usually 2 or $2\frac{1}{2}$ parts by measure of sand to one of slaked lime in paste, or 5 to 6 parts of sand to one of unslaked lime. As lime tends to air-slake, it should be used when recently burned.

Some limes slake rapidly and completely; other limes have lumps which slake slowly and should be allowed time to combine with the water. It is generally considered that lime mortar improves by standing, and that mortar intended for plastering should be made several days before it is used. Small unslaked fragments in the plaster will swell later and crack the finished surface. The lime paste is sometimes strained to remove such lumps.

Lime mortar hardens by the drying out of part of the water which it contains, and by the slow absorption of carbon dioxide from the air. It thus passes back by degrees to a crystallized calcium carbonate surrounding the particles of sand: $Ca(OH)_2 + CO_2 = CaCO_3 + H_2O$. Dampness of the mortar is favorable to the attainment of this result, and the mortar in a brick wall which has been kept damp for some time will harden better than where the wall is dry. Dry, porous bricks absorb rapidly, and almost completely, the water from the mortar, and reduce it to a powder or friable mass which will not harden satisfactorily. Hence bricks should be well wetted before they are laid.

Lime mortar in the interior of a very thick wall may not harden for a long time, if at all, and hence should not be used in such a place. Slaked lime placed under water will not harden, as may be proved by experiment. In both cases such inaction is due to the exclusion of the carbon dioxide. Lime mortar should never be used in wet foundations.

Plaster for interior walls is lime mortar. Hair is added to the mortar for the first coat, so that the portion which is forced through the spaces between the laths and is clinched at the back may have sufficient tenacity to hold the plaster on the walls and ceiling.

**46. Natural Cement.**—Natural cement is made by burning almost to vitrification a rock which contains lime, silica, and alumina, that is, one which may be considered a mixture of a limestone and a clay rock. The carbon dioxide, moisture, and water of crystallization are expelled by burning. The hard fragments must then be ground to powder, the finer the better. If the rock contains the several ingredients in proper proportions, the addition of water to the powder makes a plastic mass which hardens or *sets* by crystallization. This setting may begin in a few minutes or half an hour. The hardening, the tensile and compressive strengths increase rapidly at first, and at a decreasing rate for months.

As access of air is not required for the setting of cement, the reaction taking place when water is added to the dry powder, cement mortar is used invariably under water and in wet places. It makes stronger work than lime mortar, and is generally used by engineers for stone masonry. Its greater cost than that of lime is due to the necessity of grinding the hard clinker; while lime falls to powder when wet. The proportion of sand is 1, 2, or 3 to one of cement, according to the strength desired, 2 to 1 being a common ratio for good work. The sand and cement are mixed dry and then wetted, in small quantities, to be used at once.

The addition of brick-dust from well-burned bricks to lime mortar will make the latter act somewhat like cement, or become hydraulic, as it is called. Volcanic earth has been used in the same way.

**47. Portland Cement.**—If the statement made as to the composition of cement is correct, it should be possible to make a mixture of chalk, lime or marl, and clay in proper proportions for cement, and the product ought to be more uniform in composition and characteristics than that from the natural rock. Such is the case, and in practice about three parts of carbonate of lime are intimately mixed with one part of clay and burned in kilns. During the burning the combined water and carbon dioxide are driven off and various compounds are formed of which tricalcium silicate ($3CaO, SiO_2$) is the most important, as it is the principal active element and constitutes the greater part of hydraulic cements. The resulting clinker is ground to an impalpable powder which forms the Portland cement of commerce. Fine grinding is essential, as it has been shown that the coarser particles of the cement are nearly inert.

Upon mixing the cement with water, the soluble salts dissolve and crystallize, that is, the cement *sets*. The water does not dry out during hardening as in lime mortar, but combines with some of the salts as water of crystallization. This crystallization takes place more slowly in the Portland than in the natural cements, but after the Portland cement has set it is much harder and stronger than natural cement. The slower-setting cement mortars are likely to show a greater strength some months or years after use than do the quick-setting ones, which attain considerable strength very soon, but afterwards gain but little.

**48. Cement Specifications.**—The following specifications of cement are reasonable:

## NATURAL CEMENT.

*Specific gravity:* not less than 2.8.

*Fineness:* 90 per cent. to pass a sieve of 10,000 meshes per square inch.

*Setting:* initial set in not less than fifteen minutes; final set in not more than four hours.

*Soundness:* thin pats of neat cement kept in air or in water shall remain sound and show no cracks.

*Tensile strength:* briquettes one inch square in cross-section shall develop after setting one day in air and the remaining time in water:

```
Neat..............  7 days...............100 lb.
 "   ............. 28   "  ...............200 lb.
1 cement, 1 sand..  7   "  ............... 60 lb.
 "         "   ..28   "  ...............150 lb.
```

## PORTLAND CEMENT.

*Specific gravity:* not less than 3.10.

*Fineness:* 92 per cent. to pass a sieve of 10,000 meshes per square inch.

*Setting:* initial set in not less than thirty minutes; final set in not more than ten hours. (If a quick-setting cement is desired for special work, the time of setting may be shortened and the requirements for tensile strength reduced.)

*Soundness:* a thin pat of neat cement kept in air 28 days shall not crack; another pat allowed to set and then boiled for five hours shall remain sound.

*Tensile strength:* briquettes, as for natural cement:

```
Neat ............  7 days...............450 lb.
 "   ............ 28   "  ...............550 lb.
1 cement, 3 sand..  7   "  ...............150 lb.
 "         "   ..28   "  ...............200 lb.
```

**49. Concrete.**—Concrete is a mixture of cement mortar (cement and sand) with gravel and broken stone, the materials being so proportioned and thoroughly mixed that the gravel fills the spaces among the broken stone; the sand fills the spaces in the gravel; and the cement is rather more than sufficient to fill the interstices of the sand, coating all, and cementing the mass into a solid which possesses in time as much strength as many rocks. It is used in foundations, floors, walls, and for complete structures. The broken stone is usually required to be small enough to pass through a 2-in. or $2\frac{1}{2}$-in. ring. The stone is sometimes omitted.

To ascertain the proportions for mixing, fill a box or barrel with broken stone shaken down, and count the buckets of water required to fill the spaces; then empty the barrel, put in the above number of buckets of gravel, and count the buckets of water needed to fill the interstices of the gravel; repeat the operation with that number of buckets of sand, and use an amount of cement a little more than sufficient to fill the spaces in the sand. If the gravel is sandy, screen it before using, in order to keep the proportions true. A very common rule for mixing is one part cement, three parts sand, and five parts broken stone or pebbles, all by measure.

The ingredients are mixed dry, then water is added and the mass is mixed again, after which it is deposited in forms in layers 6 or 8 inches thick. Experience has shown that a mixture wet enough to flow makes a denser concrete than a dry mixture, especially if the mass cannot be thoroughly tamped.

**50. Paint.**—When a film of *linseed-oil*, which is pressed from flaxseed, is spread on a surface it slowly becomes solid, tough, and leathery by the absorption of oxygen from the air. In order that the film may solidify more rapidly the *raw* oil may be prepared by heating and adding driers, oxides of lead and manganese, which aid the oxidation; oil treated in this way is called *boiled* oil. Driers should be used sparingly, as they lessen the durability of paint. An oil-film is somewhat porous and rather soft, hence its protective and wearing qualities can be improved by the addition of some finely ground *pigment* to fill the pores and make the film harder and thicker. Most pigments are inert. Paint, then, consists of linseed-oil, a pigment, and a drier. Varnish is sometimes added to make the paint glossy and harder, or turpentine may be used to thin it.

*Varnishes* are made by melting resin (resins are vegetable gums, either fossil or recent), combining it with linseed-oil, and

## MATERIALS.

thinning with turpentine. They harden by the evaporation of the turpentine and the oxidation of the oil and resin. The addition of a pigment to varnish makes enamel- or varnish-paint.

For painting on wood white lead, the carbonate, and white zinc, an oxide, are pigments extensively used. Iron oxide is largely used on both wood and steel. Red lead, an oxide, and graphite are pigments used on steel. Red lead acts differently from other pigments in that it unites with the oil, and the mixture hardens even if the air is excluded, so that red-lead paint must be mixed as used. Lampblack is often mixed with other pigments to advantage, or it is sometimes used alone.

As paint is used to form a protective coating, it should not be brushed out too thin, but as heavy a coat as will dry uniformly should be applied. Wood should be given a priming coat of raw linseed-oil, so that the wood shall not absorb the oil from the first coat of paint and leave the pigment without binder. In applying paint to steel-work it is essential for good work that the paint be spread on the clean, bright metal. Rust and mill-scale must be removed before painting if the coat is expected to last. As mill-scale can be removed only by the sand-blast or by pickling in acid, steel is seldom thoroughly cleaned in practice. If paint is applied to rusty iron, the rusting will go on progressively under the paint. Painting should never be done in wet or frosty weather.

## CHAPTER III.

### BEAMS.

**51. Beams: Reactions.**—A beam may be defined to be a piece of a structure, or the structure itself as a whole, subjected to transverse forces and bent by them. If the given forces do not act at right angles to the axis or centre line of the piece, their components in the direction of the axis cause tension or compression, to be found separately and provided for; the normal or transverse components alone produce the beam action or bending.

As all trusses are skeleton beams, the same general principles apply to their analysis, and a careful study of beams will throw much light on truss action.

Certain forces are usually given in amount and location on a beam or assumed. Such are the loads concentrated at points or distributed over given distances, and due to the action of gravity; the pressure arising from wind, water, or earth; or the action of other abutting pieces.

It is necessary, in the first place, to satisfy the requirements of equilibrium, that the sum of the transverse forces shall equal zero and that the sum of their moments about any point shall also equal zero. This result is accomplished by finding the magnitudes and direction of the forces required at certain given points, called the points of support, to produce equilibrium. The supporting forces or *reactions*, exerted by the points of support *against* the beam, are two or more, except in the rare case where the beam is exactly balanced on one point of support. For cases where the reactions number more than two, see § 109.

**52. Beam Supported at Two Points. Reactions.**—The simplest and most generally applicable method for finding one

BEAMS. 39

of the two unknown reactions is to find the sum of the moments of the given forces about one of the points of support, and to equate this sum with the moment of the other reaction about the same point of support. Hence, divide the sum of the moments of the given external forces about one of the points of support by the distance between the two points of support, usually called the *span*, to find the reaction at the other point of support. The direction of this reaction is determined by the sign of its moment, as required for equilibrium. The amount of the other reaction is usually obtained by subtracting the one first found from the total given load.

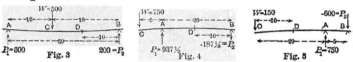

Fig. 3   Fig. 4   Fig. 5

Thus, in the three cases sketched, $P_1 = W\dfrac{CB}{AB}$; $P_2 = W - P_1$.

*Examples.*—Fig. 3. If $W = 500$ lb., A B = 30 ft., and B C = 18 ft.; $P_1 = \dfrac{500 \cdot 18}{30} = 300$ lb., $P_2 = 500 - 300 = 200$ lb.

Fig. 4. If $W = 750$ lb., A B = 20 ft., and A C = 5 ft., $P_1 = \dfrac{750 \cdot 25}{20} = 937\frac{1}{2}$ lb., and $P_2 = 750 - 937\frac{1}{2} = -187\frac{1}{2}$ lb.

Fig. 5. If $W = 150$ lb., A C = 20 ft., and A B = 5 ft., $P_1 = \dfrac{150 \cdot 25}{5} = 750$ lb., and $P_2 = 150 - 750 = -600$ lb. Note the magnitude of $P_1$ and $P_2$ as compared with $W$ when the distance between $P_1$ and $P_2$ is small. Such is often the case when the beam is built into a wall.

Where the load is distributed at a known rate over a certain length of the beam, the resultant load and the distance from its point of application to the point of support may be conveniently used.

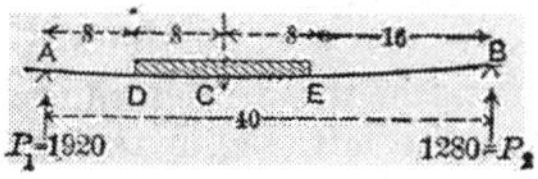

Fig. 6.

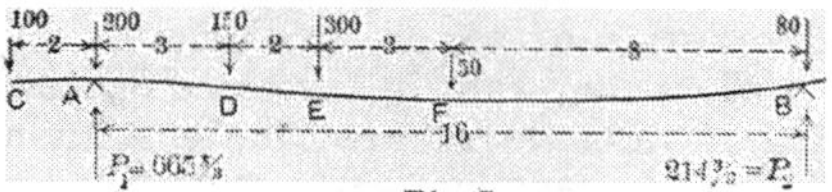

*Example.*—Fig. 6. If A B = 40 ft., A D = 8 ft., D E = 16 ft., and the load on D E is 200 lb. per ft., $W = 3{,}200$ lb., and C B = 24 ft.

Therefore $P_1 = \dfrac{3{,}200 \cdot 24}{40} = 1{,}920$ lb., and $P_2 = 3{,}200 - 1{,}920 = 1{,}280$ lb.

If several weights are given in position and magnitude, the same process for finding the reactions, or forces exerted by the points of support against the beam, is applicable.

*Examples.*—In Fig. 7, $P_1 = (100 \cdot 18 + 200 \cdot 16 + 150 \cdot 13 + 300 \cdot 11 + 50 \cdot 8 + 80 \cdot 0) \div 16 = 665\frac{5}{8}$ lb. $P_2 = 880 - 665\frac{5}{8} = 214\frac{3}{8}$ lb. The work can

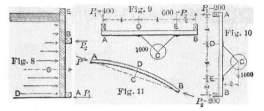

be checked by taking moments about A to find $P_2$, the moment $100 \cdot 2$ then being negative.

If the depth of water against a bulkhead, Fig. 8, is 9 ft., and the distance between A and B, the points of support, is 6 ft., A being at the bottom, the unit water pressure at A will be $9 \times 62.5 = 562.5$ lb. which may be represented by A D, and at other points will vary with the depth below the surface, or as the ordinates from E A to the inclined line E D. Hence the total pressure on E A, for a strip 1 ft. in horizontal width, will be $562.5 \times 9 \div 2 = 2{,}531\frac{1}{4}$ lb., and the resultant pressure will act at C, distant $\frac{1}{3}$ A E, or 3 ft. from A. $P_2 = 2{,}531\frac{1}{4} \times 3 \div 6 = 1{,}265.6$ lb., and $P_1 = 2{,}531.2 - 1{,}265.6 = 1{,}265.6$ lb., a result that might have been anticipated, from the fact that the resultant pressure here passes midway between A and B.

Let 1,000 lb. be the weight of pulley and shaft attached by a hanger to the points D and E, Fig. 9. Let the beam A B = 10 ft., A D = 4 ft., D E = 4 ft., E B = 2 ft.; and let C be 2 ft. away from the beam. As the beam is horizontal, $P_1 = 1{,}000 \times 4 \div 10 = 400$ lb.; $P_2 = 1{,}000 - 400 = 600$ lb., and both act upwards. The 1,000 lb. at C causes two vertical downward forces on the beam, each 500 lb., at D and E. There is also compression of 500 lb. in D E.

When the beam is vertical, Fig. 10, by moments, as before, about B, $P_1 = 1{,}000 \cdot 2 \div 10 = 200$ lb. at A acting to the left, being tension or a negative reaction. By moments about A, $P_2 = 1{,}000 \cdot 2 \div 10 = 200$ at B, acting to the right. Or $P_1 + P_2 = 0$; $\therefore P_1 = -P_2$. By similar moments, the 1,000 lb. at C causes two equal and opposite horizontal forces on the beam at D and E, of 500 lb. each, that at D being tension on the connection, or acting towards the right, and that at E acting in the opposite direction. These two forces make a couple balanced by the couple $P_1P_2$. The weight 1,000 lb. multiplied by its arm 2 ft. is balanced by the opposing horizontal forces at D and E, 4 ft. apart. There remains a vertical force of 1,000 lb. in A B, which may all be resisted by the point B, when the compression in D E = 500 lb. and in E B = 1,000 lb.; or all by the point A, when the tension in D E = 500 lb. and in D A = 1,000 lb.; or part may be resisted at A, and the rest at B, the distribution being uncertain. This longitudinal force may be disregarded in discussing the beam, as may the tension or compression in the hanger arms themselves.

**53. Bending Moments.**—If an imaginary plane of section is passed through any point in a beam, the sum of the moments of all the external forces on one side of that section, *taken about a point in the section*, must be exactly equal and opposite to the sum of the moments of all the external forces on the other side of that section, taken about the same point. If not, the beam would revolve in the plane of the forces. The moment on the left side of the section tends to make that portion of the beam rotate in one direction about the point of section, and the equal moment on the right side of the section tends to make the right segment rotate in the opposite direction. These two moments cause resistances in the interior of the beam at the section (which stresses will be discussed under *resisting moment*), with the result that the beam is bent to a slight degree. Either resultant moment on one side of a plane of section, about the section, is called the *bending moment* at that point, usually denoted by $M$, and is considered *positive* when it makes the beam *concave* on the upper side. Ordinary beams, supported at the ends and carrying loads, have positive bending moments.

If upward reactions are positive, weights must be taken as negative and their sign regarded in writing moments.

*Examples.*—Section at D, Fig. 3, 10 ft. from B. On the left of D, and about D, $P_1$ (=300)·20−500·8=2,000 ft.-lb., positive bending moment at D. Or, about D, on the right side of the section; $P_2(=200)\cdot 10 = 2,000$ ft.-lb., positive bending moment at D. Usually compute the simpler one.

Section at A, Fig. 4, $W\cdot C\, A = -750\cdot 5 = -3,750$ ft.-lb. negative bending moment at A, tending to make the beam convex on the upper side. At D, 10 ft. from B, $M = -P_2\cdot 10 = -187\frac{1}{2}\cdot 10 = -1,875$ ft.-lb., negative because $P_2$ is negative. At A, taking moments on the right of and about A, $M = -187\frac{1}{2} \cdot 20 = -3,750$ ft.-lb., as first obtained. This beam has negative bending moments at all points.

In Fig. 5, $M$ at D is $-150 \cdot 10 = -1,500$ ft.-lb. It is evident that the bending moments at all points between C and A can be found found without knowing the reactions. If this beam is built into a wall, the points of application of $P_1$ and $P_2$ are uncertain, as the pressures at A and B are distributed over more or less of the distance that the beam is embedded. The maximum $M$ is at A, and is $-150 \cdot 20 = -3,000$ ft.-lb. It is evident that the longer A B is, the smaller the reactions are, and hence the greater the security.

In Fig. 6, the bending moment at C will be $P_1 \cdot$ A C $-$ weight on D C $\cdot \frac{1}{2}$D C $= 1,920 \cdot 16 - 200 \cdot 8 \cdot 4 = 24,320$ ft.-lb. At E, $M = 1,280 \cdot 16 = 20,480$ ft.-lb.

In Fig. 7, the bending moments at the several points of application of the weights, taking moments of all the external forces on the left of each section about the section, will be—

At C, $M = -100 \cdot 0 = 0$.
At A, $M = -100 \cdot 2 = -200$ ft.-lb.
At D, $M = -100 \cdot 5 + (665\frac{5}{8} - 200) \cdot 3 = 896\frac{7}{8}$ ft.-lb.
At E, $M = -100 \cdot 7 + 465\frac{5}{8} \cdot 5 - 150 \cdot 2 = 1,328\frac{1}{8}$ ft.-lb.
At F, $M = -100 \cdot 10 + 465\frac{5}{8} \cdot 8 - 150 \cdot 5 - 300 \cdot 3 = 1,075$ ft.-lb.
And, at B, $M$ will be zero. $M$ *max*. occurs at E.

Do not assume that the maximum bending moment will be found at the point of application of the resultant of the load. The method for finding the point or points of maximum bending moment will be shown later.

The moments on the right portion of the beam may be more easily found by taking moments on the right side of any section. Thus at F, $M = (P_2 - 80) \cdot 8 = (214\frac{3}{8} - 80) \cdot 8 = 1,075$ ft.-lb. Find the bending moment at the middle of E F. $1,201\frac{9}{16}$ ft.-lb.

In Fig. 8, the bending moment at section C of the piece A E may be found by considering the portion above C. As the unit pressure at C is $6 \times 62\frac{1}{2}$ lb. $= 375$ lb. per sq. ft., $M$ at C $= P_2 (= 1,265.6) \cdot 3 - (375 \times 6 \div 2) \cdot 6 \div 3 = 1,546.8$ ft.-lb. At the section B, $M = -(3 \times 62\frac{1}{2} \times 3 \div 2) \times 1 = -281\frac{1}{4}$ ft.-lb.

In Fig. 9, as $P_1 = 400$ lb., $P_2 = 600$ lb., vertical forces at D and E are each 500 lb.; $M$ at D $= 1,600$ ft.-lb.; $M$ at E $= 1,200$ ft.-lb.

In Fig. 10, as $P_1 = -200$ lb. $= -P_2$, and the horizontal forces at D and E are $\pm$ 500 lb.; $M$ at D $= -800$ ft.-lb.; $M$ at E $= +400$ ft.-lb. The beam will be concave on the left side at D and convex at E. The curvature must change between D and E, where $M = 0$. Let this point be distant $x$ from B. Then $200 \cdot x - 500(x-2) = 0$; $\therefore x = 3\frac{1}{3}$ ft.

## BEAMS

The curved piece A B, Fig. 11, with equal and opposite forces applied in the line connecting its ends, will experience a bending moment at any point D, equal to $P \cdot C D$, this ordinate being perpendicular to the chord.

**54. Shearing Forces.**—In Fig. 3, of the 500 lb. at C, 300 lb. goes to A and 200 lb. to B. Any vertical section between A and C must therefore have 300 lb. acting vertically in it. On the left of such a section there will be 300 lb. from $P_1$ acting upwards, and on the right of the same section there will be 300 lb., coming from $W$, acting downwards. These two forces, acting in opposite directions on the two sides of the imaginary section, tend to cut the beam off, as would a pair of shears, and either of these two opposite forces is called the shearing force at the section, or simply the *shear*. When acting upwards on the left side of the section (and downwards on the right side), it is called *positive* shear. When the reverse is the case the shear will be negative.

*Examples.*—In Fig. 7, where a number of forces are applied to a beam, there must be found at any section between C and A a shear of $-100$ lb.; between A and D the shear will be $-100 + 665\frac{2}{3} - 200 = +365\frac{2}{3}$ lb.; between D and E the shear will decrease to $365\frac{2}{3} - 150 = 215\frac{2}{3}$ lb.; on passing E the shear will change sign, being $215\frac{2}{3} - 300 = -84\frac{1}{3}$ lb.; between F and B it will be $-84\frac{1}{3} - 50 = -134\frac{1}{3}$ lb.; and on passing B, it becomes zero, a check on the accuracy of the several calculations.

In Fig. 8, the shear just above the support $B = 3 \times 62\frac{1}{2} \times 3 \div 2 = 281\frac{1}{4}$ lb.; just below the point B the shear is $281\frac{1}{4} - 1,265.6 = -984.4$ lb.; and just above A it is 1,265.6 lb. The signs used imply that the left side of A E corresponds to the upper side of an ordinary beam. As the shear is positive above A and negative below B, it changes sign at some intermediate point. Find that point.

In Fig. 9, the shear anywhere between A and D is $+400$ lb.; at all points between D and E it is $400 - 500 = -100$ lb.; and between E and B is $-600$ lb. The shear changes sign at D.

In Fig. 10, the shear on any horizontal plane of section between B and E is $-200$ lb.; betwen E and D is $-200 + 500 = +300$ lb.; and between D and A is $+300 - 500 = -200$ lb. The shear changes sign at both E and D.

**55. Summary.**—To repeat:—The *shearing force* at any normal section of a beam may therefore be defined to be *the algebraic sum of all the transverse forces on one side of the section.*

When this sum or resulting force acts upward on the left of the section, call it positive; when downward, negative.

The *bending moment* at any right or normal section of a beam may be stated to be *the algebraic sum of the moments of all the transverse forces on one side of the section, taken about the centre of gravity of the section as axis.* When this sum or resulting moment is right-handed or clockwise on the left of and about the section, call it positive. A positive moment tends to make the beam concave on what is usually the upper side.

By a proposition in mechanics, any force which acts at a given distance from a given point is equivalent to the same force at the point and a moment made up of the force and the perpendicular from the point to the line of action of the force. Then in Fig. 7, if a section plane is passed anywhere, as between D and E, the resultant force on the left, which is the algebraic sum of the given forces on the left of the section, is the *shear* at the section; and this resultant, multiplied by its arm or distance from the point in D E, giving a moment which is the algebraic sum of the moments of the several forces on the left of and about the point, is the *bending moment* at the section.

It is also evident that the resulting action at any section is the sum of the several component actions; and hence that different loads may be discussed separately and their effects at any point added algebraically, if they can occur simultaneously. Thus the shears and bending moments arising from the weight of a beam itself may be determined, and to them may be *added* the shears and bending moments at the same points from other weights imposed on the beam.

The numerous examples already given show that formulas are not needed for solving problems in beams, and the student will do well to accustom himself to using the data directly. Formulas, however, will now be derived, which will sometimes be convenient for use, and from which may be deduced certain serviceable relationships.

**56. Bending Moment a Maximum where the Shear Changes Sign.**—If a beam weighing $w$ per unit of length is supported at each end and carries a system of loads any one of which is distant $a$ from the left support, the shear at a section distant $x$ from the left support is

$$F_x = P_1 - \Sigma_0^x W - wx,$$

and the bending moment at the same section is

$$M_x = P_1 x - \Sigma_0^x W(x-a) - \tfrac{1}{2}wx^2.$$

If the beam is a *cantilever*, that is, a beam fixed in position at the right end and unsupported at the left, the same equation will apply when $P_1$ becomes zero. It is seen by comparing the equations above that $F$ is always the first derivative of $M$, or

$$\frac{dM_x}{dx} = F_x.$$

Hence, according to the rule for determining maxima and minima, the bending moment is always a maximum (or minimum) at the place where the shear is zero or changes in sign. This criterion is easily applied to locate the points of $M$ maximum. Pass along the beam from the left (or right) until as much load is on the left (or right) of the section as will neutralize $P_1$ (or $P_2$) and the point of $M$ *max.* is found. Its value can then be computed. If the weight at a certain point is more than enough to reduce $F$ to zero, $F$ changes sign in passing that point, and hence $M$ *max.* occurs there.

For a beam fixed at one end only, $F$ changes sign in passing $P_1$, and hence $M$ *max.* is found at the wall.

*Examples.*—$M$ *max.* occurs in Fig. 3, at C; in Fig. 4, at A; in Fig. 6, at 17.6 ft. from A; in Fig. 7, at A, and again at E; in Fig. 8, at B, and again at a distance $x$ from E such that $62\tfrac{1}{2}x \cdot \tfrac{1}{2}x = 1,265\tfrac{5}{8}$; $\therefore x = \sqrt{40.5} = 6.36$ ft.; in Fig. 9, at D; and in Fig. 10, at D and again at E. The bending moments which may not have been found at some of these points can now be computed.

The reader who is familiar with graphics can draw the equilibrium or bending-moment polygons or curves, and the shear diagrams, and notice the same relation in them.

46               STRUCTURAL MECHANICS.

The *unit load* may also be considered as the derivative of the shear; $F$ therefore has maximum (or minimum) values where the external forces change in sign.

The origin of coordinates may be arbitrarily taken at any point in the length of the beam and general expressions may be written. If $-w$ is the unit load and is constant,

$$F_x = -\int w dx = F_0 - wx,$$
$$M_x = \int F_x dx = M_0 + F_0 x - \tfrac{1}{2} w x^2,$$

in which $F_0$ and $M_0$ are the constants of integration, the values of $F$ and $M$ at the origin. Thus in the beam above, the bending moment at a second section distant $c$ from the first is

$$M_{x+c} = M_x + F_x c - \Sigma_x^{x+c} W(x-a+c) - \tfrac{1}{2} w c^2.$$

The same expression is easily derived by substituting $x+c$ for $x$ in the original equation.

*Example.*—In Fig. 7, $M$ at D $= +896\tfrac{7}{8}$ ft.-lb. $F$ between D and E $= +215\tfrac{5}{8}$ lb. $M$ at E $= +896\tfrac{7}{8} + 215\tfrac{5}{8} \times 2 = 1{,}328\tfrac{1}{8}$ ft.-lb. as in § 53. $M$ midway between E and F $= 896\tfrac{7}{8} + 215\tfrac{5}{8} \times 3\tfrac{1}{2} - 300 \times 1\tfrac{1}{2} = 1{,}201\tfrac{7}{16}$ ft.-lb.

**57. Working Formulas.**—The bending moments and shears for a number of simple cases of common occurrence are given below. In general the bending moment and the shear vary from point to point along a beam, and they may be conveniently represented on a diagram by ordinates whose lengths represent the values of those quantities. In the accompanying figures the upper diagram is one of bending moments and the lower is one of shears. Positive values are laid off above the base line and negative below. The student who has followed the examples of the preceding sections should have no difficulty in computing the ordinates given in Figs. 12 to 15.

Figs. 12 and 13 represent cantilever beams, one carrying a load of $W$ at the end, the other carrying a uniformly distributed

## BEAMS. 47

load of $w$ per unit of length. The bending moment at any section distant $x$ from the free end of the beam of Fig. 13 is $M = -wx \cdot \tfrac{1}{2}x = -\tfrac{1}{2}wx^2$ and the bending-moment diagram is therefore a parabola.

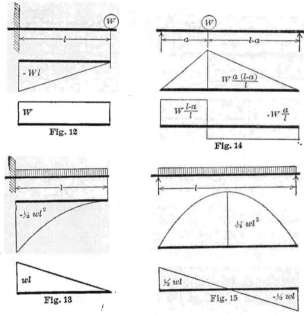

Fig. 12

Fig. 14

Fig. 13

Fig. 15

Fig. 14 shows a beam on two supports carrying a load $W$. The bending moment is evidently a maximum under the weight, and is equal to $W\dfrac{a(l-a)}{l}$, a quantity easily remembered as the weight into the product of the two segments divided by the span. When the load is at mid-span $M = \tfrac{1}{4}Wl$. In Fig. 15 a load of intensity $w$ per unit of length is distributed over the beam. The left reaction is $\tfrac{1}{2}wl$, and the bending moment at a section distant $x$ from the left support is

$$M = \tfrac{1}{2}wl \cdot x - wx \cdot \tfrac{1}{2}x = \tfrac{1}{2}wx(l-x).$$

The right-hand member of this equation contains $x(l-x)$, which is the product of two variables whose sum is constant; therefore the bending-moment diagram is a parabola.

The above diagrams are drawn for the applied loads alone, that is, the beams are considered to be without weight. If a beam weighing $w$ per unit of length carries a load $W$ as in Fig. 12 or 14, the bending moment or shear at any section can be found by adding the ordinates of Figs. 12 and 13 or of Figs. 14 and 15.

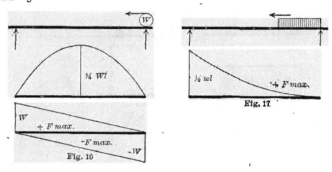

At any section distant $x$ from one end of a beam on two supports, the bending moment due to a single *moving* load is a maximum when the load is at the section, when it has a value of $M = Wx(l-x) \div l$. As this is equation of a parabola of altitude $\tfrac{1}{4}Wl$, the absolute maximum moments which can occur in the beam under a single moving load are as shown in Fig. 16. The greatest positive shear at any section occurs when the load is just to the right of that section; it is equal to the left reaction and consequently is proportional to the distance of that section from the right support, hence the locus of maximum positive shear is a straight line as shown.

If a uniform load advances continuously from the right end of the beam of Fig. 17, the positive shear at any section will increase until the load reaches the section, after which it will decrease as the load extends into the left segment. This is evident from the fact that any load placed in the right segment causes positive shear at the section, while any load in the left segment causes negative shear at the section; see Fig. 14. When the road extends a distance, $x$, from the right support the shear at the head of the load is $F = P_1 = \tfrac{1}{2}wx^2 \div l$ and the curve of max-

BEAMS. 49

imum shears is a parabola. To cause maximum negative shears the load must advance from the left. The maximum moment at any section occurs when the whole span is loaded, in which case the moments are given by Fig. 15.

**58. Position of Wheel Concentrations for Maximum Moment at any Given Point.**—Where moving loads have definite magnitudes and spacings, as is the case with the wheel weights of a locomotive, the position of the load, on a beam or girder supported at both ends, to give maximum bending moment at any given

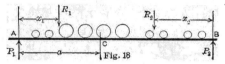

Fig. 18

section may be found as follows:—Let the given section be C, at a distance $a$ from the left abutment of a beam A B, of span $l$, Fig. 18.

Let $R_1$ be the resultant of all loads on the left of C, and acting at a distance $x_1$ from the left abutment; let $R_2$ be the resultant of all loads to the right of C, and acting at a distance $x_2$ from the right abutment. The reaction $P_1$ at A, due to $R_2$ alone, is $R_2 x_2 \div l$, and the bending moment at C, due to $R_2$ only, is $R_2 a x_2 \div l$. Similarly $P_2$, due to $R_1$, is $R_1 x_1 \div l$, and the bending moment at C, due to $R_1$ only, is $R_1(l-a)x_1 \div l$. Hence the total bending moment at C is

$$M = R_2 \frac{a x_2}{l} + R_1 \frac{(l-a)x_1}{l}.$$

If the entire system of loads is advanced a short distance $d$ to the *left*, the bending moment at C becomes

$$M' = R_2 \frac{a(x_2+d)}{l} + R_1 \frac{(l-a)(x_1-d)}{l}.$$

The *change of bending moment* due to moving the loads to the *left* is

$$M' - M = R_2 \frac{ad}{l} - R_1 \frac{(l-a)d}{l}.$$

If the loads are moved a distance, $d$, to the *right* instead of to the left, the change of bending moment is

$$M' - M = -R_2\frac{ad}{l} + R_1\frac{(l-a)d}{l}.$$

From the two values of $M'-M$ it is seen that the bending moment at C will be increased by moving the loads to the left when $R_2 a > R_1(l-a)$ and by moving to the right when $R_1(l-a) > R_2 a$. When $R_1(l-a) = R_2 a$ the bending moment cannot be increased by moving either to the right or left and is therefore a maximum. The condition may be written

$$\frac{R_1}{a} = \frac{R_2}{l-a} = \frac{R_1 + R_2}{l}.$$

$R_1 \div a$ is the average load per unit of length on the left segment and $(R_1 + R_2) \div l$ is the average load on the span, hence

The bending moment at any point of a beam carrying a system of moving loads is a maximum when the *average load on one segment is equal to the average load on the span*.

Since, for maximum bending moment at any section, a load must be at that section, place a load $W_n$ at the given point and compute the above *inequality*, first considering $W_n$ as being just to the right and then just to the left of the section. If the inequality changes sign, the position with $W_n$ at the section is one of *M max.* The value of *M max.* can be computed as in § 53. If, however, the inequality does not change sign, move the whole system until the next $W$ comes to the section, and test the inequality again.

It sometimes happens that two or more different positions of the load will satisfy the condition just explained, and, to determine the absolute *M max.*, each must be worked out numerically. When there are some $W$'s much heavier than others, *M max.* is likely to occur under some one of them. When other loads are brought on at the right, or pass off at the left, they must not be overlooked.

**59. Position of Wheel Concentrations for Maximum Shear at any Given Point.**—The shear at point C in the beam or girder of Fig. 19, as the load comes on at the right end, will increase until the first wheel $W_1$ reaches C. When that wheel passes C, the shear at that point suddenly diminishes by $W_1$, and then again gradually increases, until $W_2$ reaches C. Let $R$ be the sum of all loads on the span when $W_1$ is at C, and $x$ the distance from the centre of gravity of the loads to the right point of support

BEAMS.    51

B. The shear at C will be $P_1 = Rx \div l$. If the train moves to the left a distance $b$, the space between $W_1$ and $W_2$, so that $W_2$ has just reached C, the shear at C will be $R(x+b) \div l - W_1$, plus a small quantity $k$ which is the increase in $P_1$, due to any additional loads which may have come on the span during this advance of the train. The shear at C will therefore be

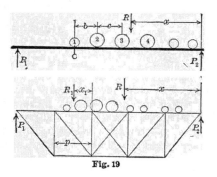

Fig. 19

increased by moving up $W_2$, if $Rb \div l + k > W_1$, or (as $k$ can often be neglected) if

$$R\frac{b}{l} > W_1 \quad \text{or} \quad \frac{R}{l} > \frac{W_1}{b}.$$

Hence move up the next load when the average load per foot on the span is greater than the load on the left divided by the distance between $W_1$ and $W_2$.

Similarly, $W_3$ should be moved to C if $\dfrac{R'c}{l} > W_2$ or $\dfrac{R'}{l} > \dfrac{W_2}{c}$, $R'$ being the sum of the loads on the span when $W_2$ is at C, and $c$ the distance between $W_2$ and $W_3$.

It is not necessary to take account of $k$ unless the two sides of the inequality are nearly equal.

*Example.*—Span 60 ft., weights in units of 1,000 lb.

|   | 1 | 2 | 3 | 4 | 5 | 6 | 7 | 8 | 9 | 10 |
|---|---|---|---|---|---|---|---|---|---|---|
| Weights = | 8 | 15 | 15 | 15 | 15 | 9 | 9 | 9 | 9 | 8 | 15 |
| Spacing = | 8' | 6' | 4½' | 4½' | 7' | 5' | 6' | 5' | 8' | 8' |

To apply test for $M$ max. at 15 ft. from left, load advancing from right. With $W_2$ at quarter span, load on span = 104. If $W_2$ is just to the right, $\dfrac{104}{60} > \dfrac{8}{15}$ or $\dfrac{104}{4} > \dfrac{8}{1}$; if $W_2$ is just to the left, $\dfrac{104}{4} > \dfrac{23}{1}$; therefore move up $W_3$ to right of quarter span. $W_{10}$ now is on the span. $\frac{112}{4} > \frac{23}{1}$. Consider $W_3$ to be just to the left; then $\frac{112}{4} < \frac{38}{1}$, or the inequality changes with $W_3$.

$$P_1 = (8\cdot59 + 15\cdot51 + 45\cdot40\tfrac{1}{2} + 36\cdot21 + 8\cdot5) \div 60 = 64.26.$$
$$M\ max. = (64.26)15 - 8\cdot14 - 15\cdot6 = 761,900\ \text{ft.-lb.}$$

To test for $F\ max.$ at same point. Put $W_1$ at quarter span. Load on span = 95. $\frac{95}{60} > \frac{8}{8}$. Move up $W_2$; load on span now 104. $\frac{104}{60} < \frac{15}{6}$. Inequality changes. $P_1 = (8\cdot53 + 15\cdot45 + 45\cdot34\tfrac{1}{2} + 36\cdot15) \div 60 = 53.2$. $F.\ max. = P_1 - W_1 = 53.2 - 8 = 45,200$ lb.

When these locomotive wheel loads are distributed to the panel-joints of a bridge truss through the longitudinal stringers, which span the panel distance between floor-beams, the above rule is modified. A load in a panel being supported directly by the stringers is by that means carried to the joints of the truss. When the train advances from the right end until the forward wheels are in the panel under investigation, the shear in that panel is the left reaction minus such part of the loads in the panel as go to the floor-beam to the left. Let $R$ be the resultant of all the loads on the span applied at a distance $x$ from the right support, and let $R_1$ be the resultant of the loads in the panel, of length $p$, applied at a distance, $x_1$, from the right floor-beam. Then

$$F = R\frac{x}{l} - R_1\frac{x_1}{p}.$$

If the loads are moved a distance, $d$, to the left, the change of shear is

$$F' - F = R\frac{d}{l} - R_1\frac{d}{p},$$

and by an argument similar to that of the preceding section, for a maximum

$$\frac{R}{l} = \frac{R_1}{p}.$$

Hence the shear in any panel of a truss is a maximum when the *average load in the panel is equal to the average load on the span.*

**60. Absolute Maximum Bending Moment on a Beam under Moving Loads.**—When a beam or girder of uniform cross-section, such as a rolled I beam, supported at its ends, is subjected to a system of passing loads, such as an engine, heavy truck, or trolley, it generally suffices to determine that position of the system of weights which causes the absolute maximum bending moment, the section where it is found, and its amount.

In Fig. 18 let C be that section. Let $R=$ resultant of all loads on the beam and $x=$ its distance from B; $R_1=$ resultant of the loads to the left of C. The reaction at left, $P_1 = Rx \div l$; and, since the bending moment at C is to be a maximum, the shear at C must be zero, or

$$R\frac{x}{l} - R_1 = 0. \qquad \therefore \frac{R}{l} = \frac{R_1}{x}.$$

But the position of loads must also satisfy the condition of § 58, since there is to be maximum bending moment at C, and

$$\frac{R_1}{a} = \frac{R}{l}. \qquad \therefore \frac{R_1}{a} = \frac{R_1}{x}, \qquad \text{or} \qquad a = x.$$

The point of absolute maximum bending moment therefore is as far from one end as the centre of gravity of the whole load is from the other. The rule may be written: *Place the loads so that the centre of the span bisects the distance between the centre of gravity of the whole load on the span and a neighboring wheel.* If the shear changes sign under that wheel, the loads on the span are placed to cause the greatest possible bending moment.

*Example.*—Beam of span 24 ft. Two wheels 6 ft. apart, one carrying 2,000 lb. and one 4,000 lb., pass across. Centre of gravity is $2,000 \cdot 6 \div 6,000 = 2$ ft. from the heavier wheel. Then this wheel is to be placed 1 ft. from mid-span. Reaction $= 6,000 \cdot 11 \div 24 = 2,750$ lb. $M. \ max. = 2,750 \cdot 11 = 30,250$ ft.-lb.

**61. Total Tension Equals Total Compression.**—If a beam, loaded in any manner, and in equilibrium under the moments caused by the external forces, is cut perpendicularly across by an imaginary plane of section, while the right-handed and left-handed bending moments already shown to exist, § 53, continue to act, it is evident that the left and right segments of the beam can only be restrained from revolving about this section by the internal stresses exerted between the material particles contiguous to the section. These stresses must be of such signs, that is tensile and compressive; of such magnitude, provided the material does not give way; and so distributed over the cross-section, as to make a *resisting moment* just equal to the bending moment at the section. For the former is caused by the latter and balances it.

Since the moment arms of these stresses lie in the perpendicular plane of section, the components to be considered now will be normal to the section. The tangential components are caused by and balance the external shear.

As the external forces which tend to bend a beam are all transverse to it, and have no horizontal components, the internal stresses of tension and compression which are caused by the bending moment must be equal and opposite, as required for a moment or couple, and hence the *total normal internal tension* on any section *must equal the total normal compression.*

When any oblique or longitudinal external forces act on a beam, there is always found that resultant normal stress on any right section which is required to give equilibrium.

**62. Distribution of Internal Stress on any Cross-section.—** It may be convenient in the beginning to consider one segment of the beam removed, and equilibrium to be assured between the external moment tending to rotate the remaining segment and the resisting moment developed in the beam at the section, as shown in Fig. 20.

If two parallel lines near together are drawn on the side of a beam, perpendicularly to its length, before it is loaded, these lines, when the beam is loaded to any reasonable amount and bent by that loading, will still be straight, as far as can be observed from most careful examination; but they will now converge to a point known as the centre of curvature for that part of the beam.

An assumption, then, that any and all right *sections* of the beam, being *plane before flexure, are* still *plane after* the *flexure* of this beam, is reasonable. If the right sections become warped, that warping would apparently cause a cumulative endwise movement of the particles at successive sections, especially in a beam subjected to a constant maximum bending moment over a considerable portion of its span; and such a movement and

BEAMS. 55

resulting distortion of the trace of the sectional plane ought therefore to become apparent to the eye. Such a warping can be perceived in shafts, other than cylindrical, subjected to a twisting couple, but cannot be found in beams.

The lines A C and B D just referred to will be found to be farther apart at the convex side of the beam, and nearer together at the concave side, than they first were; hence a line G H, lying somewhere between A B and C D, is unchanged in length. If, in Fig. 20, a line parallel to A C is drawn through H, the extremity

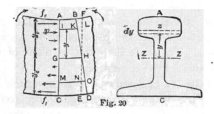
Fig. 20

of the fibre G H which has not changed in length, K L will represent the shortening which I L has undergone in its reduction to I K, and N O will represent the lengthening which M N has experienced in stretching to M O. The lengthening or shortening of the fibres, whose length was originally G H = $ds$, is directly proportional to the distance of the fibre from G H, the place of no change of length, and hence of no longitudinal or normal stress.

The diagram, Fig. 1, representing the elongation or shortening of a bar under increasing stresses, shows that, for stresses within the *elastic limit*, equal increments of lengthening and shortening are occasioned by equal increments of stress. If this beam has not been loaded so heavily as to produce a unit stress on any particle in excess of the elastic limit (and no working beam, one expected to last permanently, should be loaded to excess), the longitudinal unit stresses between the particles will vary as the lengthening and shortening of these fibres, that is, as the distance from the point of no stress. Hence, at any section, *the direct stress is uniformly varying*, with a maximum tension on the convex side and a maximum compression on the concave side.

The stresses on different forms of cross-section A C are shown in Fig. 21. The total tension on the section is always equal to the total compression.

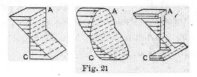

Fig. 21

**63. Neutral Axis.**—The arrows in Figs. 20, 21 may be taken to represent the unit stress at each point of the cross-section, varying as the distance from the plane of no stress, and constant in the direction $z$. To locate the point or plane of no stress or *neutral axis* for successive sections:

Let $f_c$ and $f_t$ be the unit stresses of compression and tension between the particles at the extreme edge of any section, distant $y_c$ and $y_t$ from the point of no stress. It is plain that $f_c:f_t=y_c:y_t$ from similar triangles, and that the unit stress $p$ at any point distant $y$ from the point of no stress will be

$$p=\frac{f_c}{y_c}y, \quad \text{or} \quad \frac{f_t}{y_t}y, \quad \text{or, in general} \frac{f}{y_1}y,$$

from a similar proportion.

If $zdy$ is the area of the strip on which the unit stress $p$ is exerted, $z$ being the variable coordinate at right angles to $x$ and $y$, the total *force* on $zdy$ will be $pzdy=\frac{f}{y_1}yzdy$, where $\frac{f}{y_1}$ is a constant, the unit stress at a unit distance.

As the total normal tension on the section is to equal the total compression, or their sum is to be zero, § 61, the condition may be written

$$\frac{f}{y_1}\int_{-y_t}^{+y_c} yzdy=0.$$

Therefore the sum of the moments $zdy \cdot y$ of the strips $zdy$ about the axis of $z$ must balance or be zero. Then the axis of $z$ or *neutral axis must pass through the centre of gravity of* a thin plate representing *the section,* and the neutral axis of any section lies in its plane, and usually in a direction *perpendicular to the plane of the applied external forces.* The axes of the successive cross-sections make up what is known as the *neutral plane* of the beam. Although there is no longitudinal or normal tension or compression at that line of the cross-section, it experiences shear, as will be shown later.

**64. Resisting Moment.**—The law of the variation of stress on the cross-section and the location of the neutral axis have been established. The resisting moment is caused by and is equal to the bending moment. The moments of all the stresses about the neutral axis Z Z is, since $p$ has the same sign as $y$, and the moments conspire,

$$M = \int (+p)zdy(+y) + \int (-p)zdy(-y) = \int_{-y_t}^{+y_c} pyzdy.$$

As $f \div y_1$ denotes the unit stress at either extreme fibre divided by its distance from the neutral axis, and $p = \dfrac{f}{y_1} y$,

$$M = \frac{f}{y_1} \int_{-y_t}^{+y_c} y^2 zdy = \frac{fI}{y_1}, \quad \text{and} \quad f = \frac{My_1}{I},$$

where $I$ represents $\int y^2 zdy$ about the axis Z Z, lying in the plane of the section, through the centre of gravity of the same and perpendicular to the plane of the external forces applied to the beam. $I$ is termed in mechanics the *moment of inertia* of a plane area, and is usually one of the principal moments of inertia of the area. The integral will be of the fourth power, involving the breadth and the cube of the depth. For moments of inertia of plane sections, see Chap. IV.

In the above expression for the resisting moment the quantity $I \div y_1$ is known as the *section modulus.* The section moduli of steel beams, angles, etc., are tabulated in the handbooks published by the various steel manufacturers, so that the resisting moment of a steel beam can be readily found by multiplying the section modulus by the working stress.

As moments of inertia of plane areas are of the fourth power, and can be represented by $n'bh^3$, where $h$ is the extreme dimension parallel to $y$, and $b$ to $z$, and as $y_1$ may be written $m'h$, the resisting moment can be represented, if $n' \div m' = n$, by

$$M = \frac{fI}{y_1} = nfbh^2,$$

$n$ being a fraction. For a rectangular section this becomes

$$M = f \cdot \frac{bh^3}{12} \div \frac{1}{2}h = \frac{1}{6}fbh^2;$$

and for a circular section

$$M = f \cdot \frac{\pi d^4}{64} \div \frac{1}{2}d = \frac{\pi}{32}fd^3 = 0.0982 fd^3.$$

*Examples.*—A timber beam $6'' \times 12''$, set on edge, with a safe unit stress of 800 lb. will safely resist a bending moment amounting to $800 \cdot 6 \cdot 12^2 \div 6 = 115,200$ in.-lb.

A round shaft, 3 in. in diameter, if $f = 12,000$ lb. will have a safe resisting moment of $12,000 \cdot 22 \cdot 3^3 \div 7 \cdot 32 = 31,820$ in.-lb.

For rectangular sections, either $b$ or $h$ is usually assumed and $h$ or $b$ then found. If the ratio $h \div l$ is fixed by the desire to secure a certain degree of stiffness (see "Deflection of Beams," Chap. VI.), the unknown quantity is $b$.

*Example.*—A wooden beam of 12 ft. span carries 3,600 lb. uniformly distributed. $M = \frac{1}{8}Wl = \frac{1}{8} \cdot 3,600 \cdot 12 \cdot 12 = 64,800$ in.-lb. If $f = 1,000$, $E = 1,400,000$, and the deflection $v$ is not to exceed $\frac{1}{600}$ of the span, from $\frac{v}{l} = \frac{5fl}{48Ey_1}$ is obtained $\frac{1}{600} = \frac{5}{48} \cdot \frac{1,000 \cdot 2 \cdot 144}{1,400,000 \cdot h}$; $\therefore h = 13$ in. Then assuming $h = 14$ in., a practicable size, $64,800 = \frac{1,000}{6} b \cdot 14^2$; and $b = 2$ in.

Economy of material apparently calls for as large a value of $h$ as possible; but the breadth $b$ must be sufficient to give lateral stiffness to the beam, or it may fail by the buckling or sidewise flexure of the compression edge, between those points where it

BEAMS. 59

is stayed laterally. The effect of loading as a beam a thin board set on edge will make clear the tendency.

When the plane of the applied forces does not pass through the axis of the beam, a twisting or torsional moment is added, which will be discussed in § 86.

**65. Limit of Application of $M = fI \div y_1$.**—The expression for the resisting moment at any section of a beam, caused by and always equal to the external bending moment at that section, is applicable only when the maximum unit stress $f$ does not exceed the unit stress at the elastic limit of the material. If $f$ exceeds that limit, a uniformly varying stress over the whole section is not found, and the neutral axis may not remain at the centre of gravity. Hence, also, the substitution of breaking weights, obtained by experiments on beams which fail, in a bending-moment formula which is then equated with $fI \div y_1$, results in values of $f$, the then so-called *modulus of rupture*, agreeing with neither the tensile nor the compressive strength of the material, and therefore of but limited value. This formula is correct for the purpose of design and construction; but its limitation should be kept in mind.

**66. The Smaller Value of $f \div y_1$ to be Used.**—Since from similar triangles $f_c \div y_c = f_t \div y_t$, it is immaterial which ratio is used for $M$ for a *given* cross-section. But, in designing a cross-section to resist a *given moment*, if $y_t$ and $y_c$ are not to be equal, another consideration has weight. A numerical example will bring out the distinction.

A beam of 24 in. span is loaded at the middle with a weight of 500 lb. $M$ *max.* will be $\frac{1}{4}Wl = 500 \cdot 6 = 3,000$ in.-lb. If the depth of the beam is 5 in., and its section is of such a form that the distance from its centre of gravity to the lower edge is 2 in., and to the upper edge is 3 in., while $I = 4$, then $3,000 = \frac{1}{2}f_t \cdot 4$ or $\frac{1}{3}f_c \cdot 4$. Hence the maximum unit tension $f_t = 1,500$ lb. per sq. in., and the maximum unit compression $f_c = 2,250$ lb. per sq. in. But if the material of the above beam must not be subjected to a unit stress greater than 2,000 lb. per sq. in., that unit stress will be found on the compression side; for 2,000 lb. per sq. in. on the tension side would be accompanied by 3,000 lb. per sq.

in. on the compression side; and a unit stress of 2,000 lb. compression is only compatible in this case with $2{,}000 \cdot \tfrac{2}{3} = 1{,}333$ lb. unit stress tension. The beam will safely carry only a moment of $2{,}000 \cdot 4 \div 3 = 2{,}667$ in.-lb.

Hence, when designing, with a maximum allowed value of $f$, and using a form of section where $y_t$ and $y_c$ differ, take that ratio of $f \div y_1$ which is the smaller. For a few materials, where $f_c$ and $f_t$ may be taken as differing in magnitude, as perhaps in cast iron, use that ratio $f_c \div y_c$ or $f_t \div y_t$ which gives the smaller value. As the elastic limit in tension and compression for a given material is usually the same, use in computations the larger value of $y_1$.

**67. Curved Beams.**—An originally curved beam, at any given cross-section made at right angles to its neutral axis, so far as the resisting stresses to bending moments are concerned, is in the same condition with an originally straight beam at a similar and equal cross-section to which the same bending moment is applied. Any definite thrust or tension at its two ends adds a moment at each right section equal to the product of the force into the perpendicular ordinate from the chord to the centre of the section, and a force, parallel to the chord, which force can be resolved into one normal to the section and a shear. Compare Fig. 11.

**68. Inclined Beams.**—A sloping beam is to be treated like a horizontal beam, so far as resisting stress produced by that component of the load which is normal to the beam is concerned. The component of the load which acts along the beam is to be considered as producing a direct thrust along the beam if taken up at the lower end; or a direct tension, if taken up at the upper end, or as divided somewhat indeterminately, if resisted at both ends. If this longitudinal force is axial, the mean unit stress $f_c$ caused by it is to be added to the stress $f_b$ of the same kind from bending moment at the section where this sum $f_c + f_b$ will be a maximum. This point can easily be found graphically. If the section of the piece is the unknown quantity, it will commonly suffice to use the value of $M$ *max.* to determine an approximation to $f_b$, and to correct the section by the resulting value of $f_c + f_b$ at the point where the sum is largest.

If the direct force at the end or ends is not applied axially,

BEAMS.                                           61

its moment at any section may augment or diminish the bending moment of the normal components of the load.

Cases of inclined beams, for a given load and inclination, are better solved directly than by the application of formulas.

*Example.*—A wooden rafter, 15 ft. long, has a horizontal projection of 12 ft., and a rise of 9 ft., and it carries a uniformly distributed load of 1,500 lb. The normal component of this load will be 1,200 lb., the component along the roof 900 lb. The maximum bending moment, at the middle, will be $\frac{1,200 \times 15 \times 12}{8} = 27,000$ in.-lb. If the safe stress is 1,000 lb., the section to carry this moment should be $\frac{1,000 b h^2}{6} =$ 27,000, or $bh^2 = 162$. If $b=3$, $h=8$ in. If the mean thrust, at the middle of the rafter, is 1,250 lb., the maximum thrust, at the bottom end, will be 1,700 lb., and the minimum thrust, at the top end, will be 800 lb. The section of maximum fibre stress will be a very little below the middle. But, if the rafter is $3'' \times 8''$, $f_b$ from bending moment will be $\frac{27,000 \cdot 6}{3 \cdot 8 \cdot 8} = 844$ lb. Also, $f_c = \frac{1,250}{24} = 52$ lb. Hence $f_c + f_b = 896$ lb., a satisfactory result, if the rafter is stayed laterally by the roof-covering or otherwise.

**69. Movement of Neutral Axis if Yield-point is Exceeded.**— If it is assumed that cross-sections of a beam still remain plane after the yield-point is passed at the extreme fibres, the stretch and shortening of the fibres at any cross-section will continue to vary with the distance from the neutral axis or plane. Suppose then that the elongation per unit of length of the outer tension fibre has attained an amount equal to O L, Fig. 1. The unit stress on that fibre will be L N. A fibre lying half-way from that edge to the neutral axis will have a unit stress K M. If the beam is rectangular, the total tension on the cross-section must be the area O M N L, O L now being the *distance from the neutral axis* of the beam *to the tension edge.* Since the total compression on the section must equal the total tension, an equal area O L' N' must be cut off by L' N' and the compression curve. The neutral axis must then divide the given depth of the beam in the ratio of O L to O L', shifting in this case towards the compression side. Had the compression curve been below the tension curve, the neutral axis would have shifted towards the convex side of the beam.

Since L N is less than L' N', the unit stress on the extreme fibre on the tension side is the less. Hence this displacement of the neutral axis favors the weaker side. If such action continued to the time of fracture, it would account for the fact that the application of the usual formula, $fI \div y_1$, to breaking moments gives a value of $f$ which lies between the ultimate tensile and compressive strengths of the material. It must be borne in mind, however, that the compression portion of the section increases in breadth and the tension portion contracts, quite materially for ductile substances, thus adding to the complication. A soft steel bar cannot be broken by flexure as a beam at a single test.

A rectangular cross-section also tends to assume the *section* shown in Fig. 22. The compressed particles in the middle of the width can move up more readily than they can laterally, making the upper surface convex as well as wider, and the particles below at the edges, being drawn or forced in, are crowded down, making the lower surface concave as well as narrower.

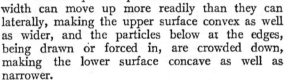

Fig. 22.

Hence the position of the neutral axis is uncertain, after the yield-point has been passed on either face; but it is probably moved towards the stronger side.

**70. Cross-section of Equal Strength.**—When a material will safely resist greater compression than tension, or the reverse, it is sometimes the custom to use such a form of cross-section that the centre of gravity lies nearer the weaker side. Cast iron is properly used in sections of this sort. See Fig. 21, section at right. Wrought-iron or steel sections are occasionally rolled or built up in a similar fashion, but the increase in width of the compression flange is then usually intended to increase its lateral stiffness.

If $f_t$ = safe unit tensile stress, and $f_c$ = safe unit compressive stress, the centre of gravity of the section must be found at such point that $y_t : y_c = f_t : f_c$, when the given safe stresses will occur simultaneously at the section. By composition, $y_t : y_c : h = f_t : f_c : f_t + f_c$, so that the centre of gravity should be distant from the bottom or top,

$$y_t = h \frac{f_t}{f_t + f_c}, \quad \text{or} \quad y_c = h \frac{f_c}{f_t + f_c}.$$

*Example.*—If $f_t$ = 3,000 lb., and $f_c$ = 9,000 lb., $y_t = h \cdot 3,000 \div 12,000 = \tfrac{1}{4}h$. If a cast-iron ⊥ section is to be used, base 10 in., thickness throughout of 1 in., and height of web $h'$, then, by moments around base,

$$\tfrac{1}{2}(h'+1) = \frac{\tfrac{1}{2}\cdot 10 + h'(1+\tfrac{1}{2}h')}{10+h'}; \therefore h' = 5 \text{ in.}, y_t = 1\tfrac{1}{2} \text{ in.};$$

$$I = \frac{10}{12} + \frac{125}{12} + 10\cdot 1 + 5\cdot 4 = \frac{165}{4}. \quad M = \frac{3{,}000\cdot 165\cdot 2}{4\cdot 3} = 82{,}500 \text{ in.-lb.},$$

the moment that the section will carry.

**71. Beam of Uniform Strength.**—As has been shown in § 64, the resisting moment may be put into the form $M = nfbh^2$, where $n$ is a numerical factor depending on the form of cross-section. If, then, for a given load, $bh^2$ be varied at successive cross-sections to correspond with the variation of the external bending moment, the unit stress on the extreme fibre will be constant; the beam will be equally strong at all sections, except against shear; and there will be no waste of material for a given type of cross-section, provided material is not wasted in shaping.

Suppose, for example, that a beam is to be supported at its ends, to carry $W$ at the middle, and to be rectangular in cross-section. The bending moment at any point between one support and the middle is $\tfrac{1}{2}Wx$. Equate this value with the resisting moment. $\tfrac{1}{2}Wx = \tfrac{1}{6}fbh^2$. To make $f$ constant at all cross-sections, $bh^2$ must vary as $x$ from each end to the middle. If $h$ is constant, $b$ must vary as $x$, or the beam will be lozenge-shaped in plan and rectangular in elevation. If, on the other hand, $b$ is constant, $h^2$ must vary as $x$, and the elevation will consist of two parabolas with vertices at the ends of the beam and axis horizontal, while the plan will be rectangular.

The section need not be a rectangle. If the ratio of $b$ to $h$ is not fixed, the treatment will be like the above; but, if that ratio is fixed, as for a circular section, or other regular figure, $b = ch$, and $h^3$ must vary as the external bending moment, or, in the case above, as $x$. The cross-section of the cast-iron beam in the example of the previous section may be varied in accordance with these principles.

The following table gives the shape of beams of rectangular cross-section supported and loaded as stated.

When a beam supported at both ends carries a single moving load $W$, passing across the beam, the bending moment at the

| Beam. | M. | $bh^2$ varies as | $h^2$ Constant, $b$ varies as | $b$ Constant, $h^2$ varies as |
|---|---|---|---|---|
| Fixed at one end, $W$ at other. | $-Wx$ | $x$ | $x$, triangular plan, Fig. 23. | $x$, parabolic elevation. Fig. 24. |
| Fixed at one end, uniform load. | $-\tfrac{1}{2}wx^2$ | $x^2$ | $x^2$, parabolic plan, Fig. 25. | $x^2$, $h$ varies as $x$, triangular elevation. Fig. 26. |
| Sup't'd both ends $W$ at $a$ from end. | $W\dfrac{l-a}{l}x$ $\dfrac{Wa}{l}(l-x)$ | $x$ $l-x$ | $\left.\begin{array}{c}x\\l-x\end{array}\right\}$ triangular plan, Fig. 27. | $\left.\begin{array}{c}x\\l-x\end{array}\right\}$ parabolic elevation. Fig. 28. |
| Sup't'd both ends, uniform load. | $\tfrac{1}{2}wx(l-x)$ | $x(l-x)$ | $x(l-x)$ parabolic plan. Fig. 29. | $x(l-x)$, circular or elliptical elevation. Fig. 30. |

point $x$, where the load is at any instant, $= Wx(l-x) \div l$. Such a beam will therefore fall under the last class of the above table.

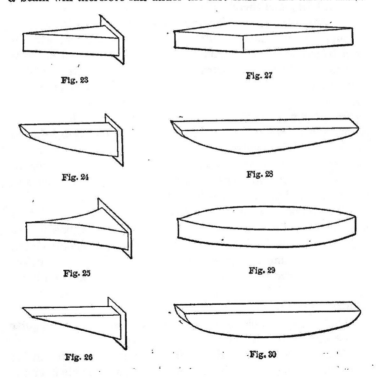

Fig. 23

Fig. 27

Fig. 24

Fig. 28

Fig. 25

Fig. 29

Fig. 26

Fig. 30

Beams which can be cast in form or built up may be made in the above outlines, if desired. Some common examples, such as brackets, girders of varying depth, walking-beams, cranks, grate-bars, etc., are more or less close approximations to such forms. Enough material must also be found at any section to resist the shear, as at the ends of beams supported at the ends.

Where a plate girder is used (see Fig. 95) with a constant depth, the cross-section of the flanges, or their thickness when their breadth is constant, will theoretically and approximately follow the fourth column of the preceding table. If the flange section is to be constant or nearly so, the depth must vary in the same way, and not as in the fifth column.

Roof- and bridge-trusses are beams of approximate uniform strength, for the different allowable unit stresses and for changing loads. The principles of this section have an influence on the choice of outline for such trusses, and the shapes of moment diagrams suggest truss forms.

**72. Distribution of Shearing Stress in the Section of a Beam, Pin, etc.**—It will be proved, in § 154, that, at any point in a body under stress, the unit shear on a pair of planes at right angles must be equal. Whatever can be proved true in regard to the unit shear on a *longitudinal* plane at any point in a beam must therefore be true of the unit shear on a *transverse* plane at the same point.

Fig. 31 represents a portion of a beam bent under any load. The existence of shear on planes parallel to E F is shown by the tendency of the layers to slide by one another upon flexure. Let the cross-section of the beam be constant. If the bending moment at section H, a point close to G, differs from that at G, there will be a shear on the transverse section, because the shear is the first derivative of the bending moment, § 56. The direct stress, here compression, on the face H F of the solid H F E G, will differ from that on the face G E, since the bending moments are different, and that difference will be balanced by a longitudinal horizontal force, or shear,

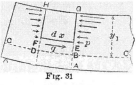

Fig. 31

on the plane F E, to oppose the tendency to displacement. If this force along the plane E F is divided by the area E F over which it is distributed, the longitudinal unit shear will be obtained. It follows from the first paragraph that the unit shear at the point E on the *transverse* section G A must be the same. It is also evident that the farther E F is taken from H G, the greater will be the difference between the total force on H F and that on G E, until the neutral axis is reached, and that the unit shear on the longitudinal plane E F must increase as E F approaches B, the neutral axis. The same thing is true if the plane is supposed to lie at different distances from the edge A. Hence, at any transverse section A G, the unit shear on a longitudinal plane is most intense at the neutral axis; and therefore the unit shear on a transverse section A G is unequally distributed, being greatest at B, the neutral axis, and diminishing to zero at A and G.

Pins and keys, and rivets which do not fit tightly in their holes, and hence are exposed to bending, have a maximum unit shear at the centre of any cross-section, and this shear must therefore be greater than the mean value, and must determine the necessary section.

To find the mathematical expression for the variation of shear on the plane A G:

O B D C is the trace of the neutral plane. $BD = EF$ sensibly $= dx$. $BE = y$, $BG = y_1$. Breadth of beam at any point $= z$, at neutral axis $= z_0$. Normal or direct unit stress at the point E on plane A G $= p$. Unit shear at E $= q$; maximum, at B, $= q_0$. $M$ and $F =$ bending moment and shearing force at section A G.

By § 64, $\quad j = \dfrac{My_1}{I} \quad$ and $\quad p = \dfrac{M}{I}y$.

The total direct stress on plane G E is

$$\int_y^{y_1} pz\,dy = \frac{M}{I}\int_y^{y_1} yz\,dy. \quad \ldots \ldots \quad (1)$$

The difference between $M$ at the section through B and $M$ at the section through D must be $F dx$, since $M = \int F dx$, by § 56.

BEAMS. 67

The horizontal force on E F is the excess of (1) for G E over its value for H F, or $\dfrac{Fdx}{I}\int_y^{y_1} yzdy$. Divide by the area $z_ydx$ of F E, over which this horizontal force acts, to find the unit shear.

$$q_y = \dfrac{F}{Iz_y}\int_y^{y_1} yzdy. \quad q_0 = \dfrac{F}{Iz_0}\int_0^{y_1} yzdy.$$

Since the mean unit shear $= F \div S$, the ratio of the maximum unit shear to the mean will be found by dividing $q_0$ by $F \div S$.

$$\dfrac{\text{Max. unit shear}}{\text{Mean unit shear}} = \dfrac{S}{Iz_0}\int_0^{y_1} yzdy = \dfrac{\int_0^{y_1} yzdy}{r^2 z_0},$$

where $r$ = radius of gyration of the cross-section, and $\int yzdy$ is, the moment of either the upper or lower part of the cross-section about the trace of the neutral plane. Hence the maximum unit shear will be:

Rectangle, $\dfrac{b\int_0^{\frac{1}{2}h} ydy}{\frac{1}{12}h^2 \cdot b} = \dfrac{12 \cdot h^2}{h^2 \cdot 8} = \dfrac{3}{2}$, or 50% greater.

Circle, $\int_0^R \dfrac{2\sqrt{(R^2-y^2)}ydy}{\frac{1}{4}R^2 \cdot 2R} = \dfrac{2 \cdot \frac{1}{3}R^3}{\frac{1}{2}R^3} = \dfrac{4}{3}$, or 33% greater.

Thin ring approximately $= 2$, or 100 per cent. greater than the mean unit shear.

For beams of variable cross-section $I$ will not be constant; but the preceding results are near enough the truth for practical purposes.

*Example.*—A cylindrical bridge-pin 3 in. diam., area 7.07 sq. in., has a shear of 50,000 lb. The maximum unit shear is $\dfrac{50,000}{7.07} \cdot \dfrac{4}{3} =$ 9,430 lb. per sq. in.

If the apparent allowable unit shear is reduced one quarter, as from 10,000 lb. to 7,500 lb., the same circular section for a pin will be obtained in designing as if the maximum unit shear were considered. For a rectangular section the apparent allowable shear should be reduced one-third.

*Example.*—A $4''\times 6''$ beam has at a certain section a shear of 2,400 lb.; the maximum unit shear on both the horizontal and the vertical plane, at the middle of the depth, is $\dfrac{2,400}{24}\cdot\dfrac{3}{2}=150$ lb. per sq. in.

As the shearing resistance along the grain of timber is much less than the shearing resistance across the grain, wooden beams which fail by shearing fracture along the grain at or near the neutral axis, at that section where the external shear is greatest. As the unit shears on two planes at right angles through a given point are always equal, the shearing strength of timber across the grain cannot be availed of, since the piece will always shear along the grain.

A rectangular timber beam of span $l$, carrying $w$ per unit of length, has a maximum fibre stress of

$$f=\frac{3}{4}\frac{wl^2}{bh^2},$$

and the greatest shearing stress along the neutral axis is

$$q_0=\frac{3}{4}\frac{wl}{bh}.$$

Dividing the first equation by the second gives the ratio between the maximum fibre stress and the maximum shear existing in the beam. If the corresponding ratio between the allowable stresses for the beam is greater than the ratio between the existing stresses, the beam is weaker in shear than in flexure; hence if $f$ and $q_0$ are working unit stresses, a rectangular beam carrying a uniformly distributed load should be designed for shear when

$$\frac{l}{h}<\frac{f}{q_0}.$$

BEAMS.                                   69

A beam carrying a single load at the centre should be designed for shear when

$$\frac{l}{h} \leq \frac{f}{2q_0}.$$

**73. Variation of Unit Shear.**—The distribution of shear on three forms of cross-section is indicated in Fig. 32, where

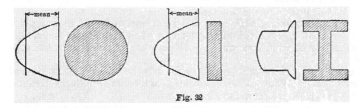

Fig. 32

the ordinates show the unit shear at corresponding points. For both the rectangle and the circle the intensity of shear varies according to the ordinates of a parabola. The curve for the I-shaped section is made up of three parabolas as shown, the intensity of shear being nearly constant over the web. For I beams of ordinary proportions and for plate girders the curve showing unit shears in the web will be much flatter than in this figure, and the maximum unit shear will differ but little from the unit shear found by dividing the total shear at a section by the cross-section of the web, as is usually done in practice.

*Examples.*—1. Three men carry a uniform timber 30 ft. long. One man holds one end of the timber; the other two support the beam on a handspike between them. Place the handspike so that each of the two shall carry ⅜ of the weight.

2. Three sections of water-pipe, each 12 ft. long, are leaded end to end. In lowering them into the trench, where shall the two slings be placed so that the joints will not be strained? Neglect the extra weight of socket.

3. Wooden floor-joists of 14 ft. span and spaced 12 in. from centre to centre are expected to carry a floor load of 80 lb. per sq. ft. If $f = 900$ lb., what is a suitable size?                          $2'' \times 10''$.

4. One of these joists comes at the side of an opening $4' \times 6'$, the load from the shorter joists, then 10 ft. long, being brought on this longer joist at 4 ft. from one end. How thick should this joist be?
                                                                4 in.

5. A cylindrical water-tank, radius 20 ft., is supported on I beams radiating from the centre. These beams are supported at one end under the centre of the tank and also on a circular girder of 15 ft. radius. They are spaced 3 ft. apart at their outer extremities. If the load is 2,000 lb. per sq. ft. of bottom of tank, find the max.+ and .−$M$ on a beam. $+29{,}600$; $-68{,}750$ ft.-lb.

6. Find $b$ and $h$ for the strongest rectangular beam that can be sawed from a round log of diameter $d$.

7. A opening 10 ft. wide, in a 16-in. brick wall, is spanned by a beam supported at its ends. The maximum load will be a triangle of brick 3 ft. high at the mid-span. If the brickwork weighs 112 lb. per cu. ft., find $M$ at the mid-span. 44,800 in.-lb.

8. In the above problem, write an expression for $M$ at any point if $w=$ weight of unit volume of load, $a=$ height of load at middle, $t=$ thickness of wall, $l=$ span, and $x=$ distance of point from the support.

9. A trolley weighing 2,000 lb. runs across a beam 6 in. wide and of 20 ft. span. What will be the elevation of a beam of uniform strength, and what its depth at middle, if $f=800$ lb.? 12 in.+

10. A round steel pin is acted upon by two forces perpendicular to its axis, a thrust of 3,000 lb. applied at 8 in. from the fixed end of the pin, and a pull of 2,000 lb. applied 6 in. from the fixed end and making an angle of 60° with the direction of the first force. Find the size of the pin, if $f=8{,}000$ lb. $M=20{,}784$ in.-lb.

11. A beam of 20 ft. span carries two wheels 6 ft. apart longitudinally, and weighing 8,000 lb. each. When they pass across the span, where and what is $M$ max.? 57,800 ft.-lb.

12. A floor-beam for a bridge spans the roadway $a$ and projects under each sidewalk $b$. If dead load per foot is $w$, live load for roadway $w'$, for sidewalk $w''$, write expressions for $+M$ max. and $-M$ max.

13. A vessel is 200 ft. long. It carries 5 tons per ft. uniformly distributed, and a central load of 300 tons. Find $M$ max. when at rest; when supported on a wave crest at bow and stern with each bearing 20 ft. long; and when supported amidships only with bearing 30 ft. long.

14. The end of a beam 6 in. wide is built into a wall 18 in. The bending moment at the wall is 600,000 in.-lb. If the top of the beam bears for 9 in. with a uniformly varying pressure and the bottom the same, what is the maximum unit compression on the bearing surface?
1,852 lb.

# CHAPTER IV.

## MOMENTS OF INERTIA OF PLANE AREAS

**74. Definitions.**—The *rectangular moment of inertia*, $I$, of a plane area about an axis lying in that plane is the sum of the products of each elementary area into the square of its distance from the axis. The plane areas whose moments of inertia are sought are commonly cross-sections of beams, and unless otherwise stated the axis about which moments are taken passes through the centre of gravity of the cross-section. The quotient of the moment of inertia by the area is the square of the *radius of gyration*, $r$. If the area be referred to rectangular axes Z and Y, and if subscripts denote the axes about which moments are taken,

$$I_z = \int\int y^2 dz dy = \int z y^2 dy = S r_z^2;$$
$$I_y = \int\int z^2 dz dy = \int z^2 y dz = S r_y^2.$$

If each elementary area be multiplied by the square of its distance from an axis perpendicular to the plane and passing through the centre of gravity, the summation gives the *polar moment of inertia*, $J$. As the distance of each elementary area from the axis is $\sqrt{z^2 + y^2}$,

$$J = \int\int (z^2 + y^2) dz dy = I_y + I_z.$$

When the term "moment of inertia" is used without qualification, the rectangular moment is meant. Moments of inertia, being the product of an area into the square of a distance, are of the fourth power and positive.

The values of $I$, $J$, and $r^2$ for some common forms of cross-section follow.

I. Rectangle, height $h$, base $b$. Fig. 33. Axis through the centre of gravity and parallel to $b$.

$$I_z = \int_{-\frac{1}{2}h}^{+\frac{1}{2}h} y^2 z\, dy = b\int_{-\frac{1}{2}h}^{+\frac{1}{2}h} y^2 dy = \left[\frac{1}{3}by^3\right]_{-\frac{1}{2}h}^{+\frac{1}{2}h} = \frac{bh^3}{24} + \frac{bh^3}{24} = \frac{bh^3}{12}.$$

$$I_y = \frac{b^3 h}{12}. \qquad r^2 = \frac{bh^3}{12} \div bh = \frac{h^2}{12}.$$

For an axis through the centre of gravity and perpendicular to the plane,

$$J = I_y + I_z = \frac{bh}{12}(b^2 + h^2), \quad \text{and} \quad r^2 = \frac{1}{12}(b^2 + h^2).$$

II. Triangle, height $h$, base $b$. Fig. 34. Axis as above and parallel to $b$.

$$h : b = \frac{2}{3}h - y : z; \qquad z = \frac{b}{h}\left(\frac{2}{3}h - y\right).$$

$$I_z = \int_{-\frac{1}{3}h}^{+\frac{2}{3}h} y^2 z\, dy = \frac{b}{h}\int_{-\frac{1}{3}h}^{+\frac{2}{3}h}\left(\frac{2}{3}h - y\right)y^2 dy = \frac{b}{h}\left[\frac{2}{9}hy^3 - \frac{1}{4}y^4\right]_{-\frac{1}{3}h}^{\frac{2}{3}h}$$

$$= \frac{b}{h}\left(\frac{16}{243} - \frac{16}{324} + \frac{2}{243} + \frac{1}{324}\right)h^4 = \frac{bh^3}{36}.$$

$$r^2 = \frac{bh^3}{36} \div \frac{bh}{2} = \frac{h^2}{18}.$$

III. Isosceles triangle, about axis of symmetry. Fig. 35. Height along axis $h$, base $b$.

$$h : \tfrac{1}{2}b = z : \tfrac{1}{2}b - y; \qquad z = h\left(1 - \frac{2y}{b}\right).$$

$$I_z = 2\int_0^{\frac{1}{2}b} h\left(1 - \frac{2y}{b}\right)y^2 dy = 2h\left[\frac{y^3}{3} - \frac{y^4}{2b}\right]_0^{\frac{1}{2}b} = 2h\left(\frac{b^3}{24} - \frac{b^3}{32}\right) = \frac{hb^3}{48}.$$

$$r^2 = \frac{b^2}{24}.$$

## MOMENTS OF INERTIA OF PLANE AREAS.    73

The sum of II and III will be the polar moment, $J$, about an axis through the centre of gravity and perpendicular to the plane.

$$J = \frac{bh}{12}\left(\frac{h^2}{3}+\frac{b^2}{4}\right); \quad r^2 = \frac{1}{6}\left(\frac{h^2}{3}+\frac{b^2}{4}\right).$$

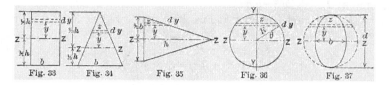

Fig. 33   Fig. 34   Fig. 35   Fig. 36   Fig. 37

IV. Circle, radius $R$, diameter $d$. Fig. 36. If $\theta =$ angle between the axis of Z and a radius drawn to the extremity of any element parallel to that axis,

$$y = R\sin\theta; \quad \tfrac{1}{2}z = R\cos\theta; \quad dy = R\cos\theta\, d\theta.$$

$$I_z = 4R^4\int_0^{\frac{\pi}{2}}\sin^2\theta\cos^2\theta\, d\theta = -4R^4\cdot\frac{1}{8}\left[\frac{1}{4}\sin 4\theta - \theta\right]_0^{\frac{\pi}{2}} = \frac{\pi R^4}{4} = \frac{\pi d^4}{64}.$$

$$r^2 = \frac{\tfrac{1}{4}\pi R^4}{\pi R^2} = \frac{1}{4}R^2 = \frac{1}{16}d^2.$$

The polar moment of inertia, $J$, may be easily written if $r'=$ variable radius,

$$J = \int_0^R r'^2 \cdot 2\pi r'\, dr' = \frac{\pi R^4}{2}. \quad r^2 = \frac{1}{2}R^2.$$

Since $I_z + I_y = J$, and $I_y = I_z$ by symmetry, $I_z = \tfrac{1}{4}\pi R^4$ as before.

V. Ellipse. Diameters $d$ and $b$. Fig. 37.

As the value of $z$ in the ellipse is to that of $z$ in the circle, as the respective horizontal diameters, or as $b$ to $d$, and as the moment of the strip $z\,dy$ varies as the breadth alone, the ellipse having horizontal diameter $b$, height $d$, gives

$$I_z = \frac{\pi d^4}{64}\cdot\frac{b}{d} = \frac{\pi b d^3}{64}; \quad r^2 = I_z \div \frac{\pi bd}{4} = \frac{d^2}{16}.$$

$$I_y = \frac{\pi d b^3}{64}; \quad J = \frac{\pi bd}{64}(d^2+b^2); \quad r^2 = \frac{1}{16}(d^2+b^2).$$

VI. The moment of inertia of a hollow section, when the areas bounded respectively by the exterior and interior perimeters have a common axis through their centres of gravity, can be found by subtracting $I$ for the latter from $I$ for the former. Thus:

Hollow rectangle, interior dimensions $b'$ and $h'$, exterior $b$ and $h$; $I_z = \frac{1}{12}(bh^3 - b'h'^3)$.

Hollow circle, interior radius $R'$, exterior radius $R$;

$$I_z = \tfrac{1}{4}\pi(R^4 - R'^4).$$
$$r^2 = \tfrac{1}{4}(R^2 + R'^2) = \tfrac{1}{16}(d^2 + d'^2).$$

The moment of inertia of a hollow ring of outside diameter $d$ and inside diameter $d'$, the ratio of $d'$ to $d$ being $n$, may be written

$$I_z = \tfrac{1}{64}\pi(d^4 - d'^4) = \tfrac{1}{64}\pi(1 - n^4)d^4.$$

When the thickness $t$ of a ring is small as compared with the diameter,

$$I_z = \tfrac{1}{8}\pi t d^3 \text{ nearly.}$$

**75. Moment of Inertia about a Parallel Axis.**—To find the moment of inertia $I'$ of a plane area about an axis Z parallel to the axis $Z_0$ through the centre of gravity and distant $c$ from it.

By definition $I' = \int (y+c)^2 z\,dy = \int y^2 z\,dy + 2c\int yz\,dy + c^2\int z\,dy$.
The first term of the second member is $I_z$, the moment of inertia about the axis through the centre of gravity; the second term has for its integral the moment of the area about its centre of gravity, which moment is zero; and the integral in the third term is the given area $S$. Hence

$$I' = I_z + c^2 S = (r^2 + c^2)S = r'^2 S.$$

*Example.*—$I_z$ for rectangle, axis parallel to $b$, is $\frac{1}{12}bh^3$. $I'$ about base $= \frac{1}{12}bh^3 + \frac{1}{4}h^2 \cdot bh = \frac{1}{3}bh^3$, and $r'^2 = \frac{1}{3}h^2$.

The reverse process is convenient for use.

$$I_z = I' - c^2 S = (r'^2 - c^2)S = r^2 S.$$

## MOMENTS OF INERTIA OF PLANE AREAS.

As the value of $I$ about an axis through the centre of gravity is the least of all $I$'s about parallel axes, it can readily be seen whether $c^2S$ is to be added or subtracted.

If the value of $I_1$ about an axis distant $c_1$ from the centre of gravity is known, and it is desired to find $I_2$ about a parallel axis distant $c_2$ from the centre of gravity, a combination of the two formulas

$$I_1 = I_z + c_1^2 S \quad \text{and} \quad I_2 = I_z + c_2^2 S$$

gives

$$I_2 = I_1 + (c_2^2 - c_1^2)S.$$

*Example.*—$I$ for a triangle about an axis through the vertex parallel to the base is easily obtained, since $z:b = y:h$.

Therefore $I$ about vertex $= \int_0^h \frac{b}{h} y^3 dy = \frac{bh^3}{4}$.

Then $I$ about base $= \frac{bh^3}{4} + \left(\frac{1}{9} - \frac{4}{9}\right) h^2 \frac{bh}{2} = \frac{bh^3}{12}$.

**76. Moments of Inertia of Shapes.**—It is frequently necessary to divide areas, such as T, I, and built iron sections, and those of irregular outline, into parts whose moments of inertia are known, each about an axis through its own centre of gravity; then to the sum of their several $I$'s add the sum of the products of each smaller area into the square of the distance from its axis to the parallel axis through the centre of gravity of the whole. This rule is an expansion of the preceding one.

$$I_z \text{ for the whole} = \Sigma I + \Sigma c^2 S.$$

Formulas for such cases are of little value. In actual computation follow the general rule.

*Example.*—Find $I$ of a $6'' \times 4'' \times \tfrac{1}{2}''$ angle about an axis parallel to shorter leg. Neglecting fillets as is customary, $S = 9.5 \times \tfrac{1}{2} = 4.75$ sq. in. Divide figure into two parts, $6 \times \tfrac{1}{2} = 3$ sq. in. and $3.5 \times \tfrac{1}{2} = 1.75$ sq. in. Distance of centre of gravity from middle of 6-in. leg is $3.5 \times 0.5 \times 2.75 \div 4.75 = 1.01$ in., which is 1.99 in. from heel.

$$I = \tfrac{1}{12} \times \tfrac{1}{2} \times 6^3 + 3 \times 1.01^2 + \tfrac{1}{12} \times 3.5 \times 0.5^3 + 1.75 \times 1.74^2 = 17.40 \text{ in.}^4$$

Or consider two rectangles, one $4'' \times 6''$, the other $3\tfrac{1}{2}'' \times 5\tfrac{1}{2}''$.

$$I = \tfrac{1}{12} \times 4 \times 6^3 + 4 \times 6 \times 1.01^2 - \tfrac{1}{12} \times 3.5 \times 5.5^3 - 3.5 \times 5.5 \times 1.26^2 = 17.40 \text{ in.}^4$$

**77. Moments of Inertia for Thin Sections.**—Values of $I$ for rolled shapes may also be approximately obtained by the following method, and, if the values given in the manufacturers' handbooks are not at hand, will prove serviceable.

Fig. 38

The moment of inertia of a thin strip or rod, Fig. 38, of length $L$ and thickness $t$, about an axis passing through one end of and making an angle $\theta$ with it, is the same as if $\tfrac{1}{3}tL$ were concentrated at the extreme end. Let $l =$ distance along strip to any particle.

$$I = \int_0^L t\,dl \cdot l^2 \sin^2\theta = \tfrac{1}{3}tL^3 \sin^2\theta = \tfrac{1}{3}tL y_1^2,$$

or one-third the area multiplied by the square of the ordinate to the extreme end.

This expression might be derived from $I$ for a rectangle, taken about one base.

If the rod is parallel to the axis, and at a distance $y_1$ from it, $I = tL \cdot y_1^2$, since all particles are equidistant from the axis.

*Example.*—Find approximate value of $I$ of angle of last example. Using centre line, angle is $5\tfrac{3}{4}'' \times 3\tfrac{3}{4}'' \times \tfrac{1}{2}''$. Distance of centre of gravity from heel is $5.75 \times 2.875 \div 9.5 = 1.74$ in.

$$I = \tfrac{1}{3} \times \tfrac{1}{2}(4.01^3 + 1.74^3) + \tfrac{1}{2} \times 3.75 \times 1.74^2 = 17.30 \text{ in.}^4$$

**78. Rotation of Axes.**—If the moments of inertia about two axes at right angles are known, the moments of inertia about axes making an angle $\alpha$ with the first may be found. In the solution the quantity $\int zy\,dS$ or $\iint zy\,dz\,dy$ occurs, which is called the *product of inertia* and represented by $Z$. Products of inertia may be either positive or negative according to the position of the axes. From Fig. 39,

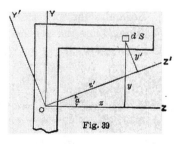

Fig. 39

$z' = z \cos\alpha + y \sin\alpha,$
$y' = y \cos\alpha - z \sin\alpha;$

## MOMENTS OF INERTIA OF PLANE AREAS.

$$I_z' = \int (y \cos \alpha - z \sin \alpha)^2 dS$$
$$= \int y^2 \cos^2 \alpha\, dS - 2\int zy \sin \alpha \cos \alpha\, dS + \int z^2 \sin^2 \alpha\, dS,$$
$$I_y' = \int (z \cos \alpha + y \sin \alpha)^2 dS$$
$$= \int z^2 \cos^2 \alpha\, dS + 2\int zy \sin \alpha \cos \alpha\, dS + \int y^2 \sin^2 \alpha\, dS$$
$$Z' = \int (z \cos \alpha + y \sin \alpha)(y \cos \alpha - z \sin \alpha) dS$$
$$= \int (y^2 - z^2) \sin \alpha \cos \alpha\, dS + \int zy \cos^2 \alpha\, dS - \int zy \sin^2 \alpha\, dS;$$

$$I_z' = I_z \cos^2 \alpha + I_y \sin^2 \alpha - 2Z \sin \alpha \cos \alpha,$$
$$I_y' = I_y \cos^2 \alpha + I_z \sin^2 \alpha + 2Z \sin \alpha \cos \alpha,$$
$$Z' = (I_z - I_y) \sin \alpha \cos \alpha + Z(\cos^2 \alpha - \sin^2 \alpha).$$

**79. Principal Axes.**—To find the maximum and minimum values of $I_z'$ and $I_y'$ differentiate with respect to $\alpha$:

$$\frac{dI_z'}{d\alpha} = -2I_z \sin \alpha \cos \alpha + 2I_y \sin \alpha \cos \alpha - 2Z \cos^2 \alpha + 2Z \sin^2 \alpha,$$

$$\frac{dI_z'}{d\alpha} = -2Z', \qquad \frac{dI_y'}{d\alpha} = +2Z'.$$

Hence maximum and minimum values of $I_z'$ and $I_y'$ occur when $Z' = 0$, and since $J = I_z + I_y = I_z' + I_y' = $ a constant, $I_y'$ is a minimum when $I_z'$ is a maximum. The axes for which $I_z'$ and $I_y'$ are maximum and minimum are called *principal axes*. *An axis of symmetry is always a principal axis* since for such an axis $Z$ must be zero, as for every positive ordinate there is a corresponding negative ordinate. If a figure has two axes of symmetry not at right angles, the moment of inertia is the same for all axes through the centre of gravity.

The angle, $\phi$, which the principal axis, A A, makes with the original Z axis may be found by making $Z' = 0$:

$$(I_z - I_y) \sin \phi \cos \phi + Z(\cos^2 \phi - \sin^2 \phi) = 0,$$
$$\tfrac{1}{2}(I_z - I_y) \sin 2\phi + Z \cos 2\phi = 0,$$

$$\tan 2\phi = \frac{2Z}{I_y - I_z}.$$

$I_A$, the moment of inertia about the A axis, is found by substituting the value of $\phi$ in the expression for $I_z{'}$:

$$I_A = I_z \cos^2\phi + I_y \sin^2\phi - Z \sin 2\phi;$$
$$\cos^2\phi = \tfrac{1}{2}(1+\cos 2\phi), \quad \sin^2\phi = \tfrac{1}{2}(1-\cos 2\phi);$$
$$I_A = \tfrac{1}{2}I_z(1+\cos 2\phi) + \tfrac{1}{2}I_y(1-\cos 2\phi) - Z \sin 2\phi$$
$$= \tfrac{1}{2}(I_y - I_z)(1-\cos 2\phi) - Z \sin 2\phi + I_z$$
$$= Z\frac{1-\cos 2\phi}{\tan 2\phi} - Z \sin 2\phi + I_z$$
$$= Z\frac{\cos 2\phi - 1}{\sin 2\phi} + I_z; \quad \tan\phi = \frac{1-\cos 2\phi}{\sin 2\phi};$$
$$I_A = I_z - Z \tan\phi.$$

Similarly the moment of inertia about axis B B at right angles to axis A A is found to be

$$I_B = I_y + Z \tan\phi.$$

**80. Oblique Loading.** — When the plane of loading on a beam does not coincide with one of the principal axes of the section the beam does not deflect in the plane of the loads and the neutral axis is oblique to that plane. In a cross-section of such a beam the stress at a point whose coordinates are $a$ and $b$ referred to the A and B axes may be found by resolving the

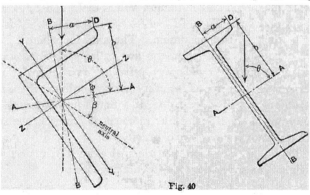

Fig. 40

bending moment at the section into its components in the direction of each principal axis, finding the stress at the point due to each

## MOMENTS OF INERTIA OF PLANE AREAS.

component and adding algebraically. Thus the stress at the point D, Fig. 40, due to a bending moment, $M$, is

$$f = \frac{(M \sin \theta)b}{I_A} + \frac{(M \cos \theta)a}{I_B}.$$

It may sometimes be necessary to find the direction of the neutral axis to determine which is the extreme or most stressed fibre. Since there is no stress at the neutral axis, if the last equation be made equal to zero, $a$ and $b$ will become coordinates of some point on the neutral axis, and the ratio between them will be the tangent of the angle, $\beta$, which the neutral axis makes with the A axis,

$$\tan \beta = \frac{b}{a} = -\frac{I_A}{I_B} \operatorname{ctn} \theta.$$

*Example.*—A $6 \times 4 \times \frac{1}{2}$-in. angle is used as a purlin on a roof of $\frac{1}{3}$ pitch ($\tan^{-1} = \frac{2}{3}$), as shown in Fig. 40. What is the maximum fibre stress due to a bending moment $M$? Centre of gravity lies 0.99 in. from back of longer leg and 1.99 in. from shorter.

$$I_x = 17.40 \text{ in.}^4, \qquad I_y = 6.27 \text{ in.}^4.$$

$Z = (\frac{1}{2} \times 4)(2-0.99)(1.99-0.25) + (\frac{1}{2} \times 5.5)(-0.99+0.25)(-3.25+1.99)$
$\phantom{Z} = +6.08 \text{ in.}^4.$

$$\tan 2\phi = 2 \times 6.08 \div (6.27 - 17.40) = -1.092,$$
$$\phi = -23° \, 46', \qquad \tan \phi = -0.440.$$

$I_A = 17.40 - 6.08(-0.440) = 20.07, \qquad I_B = 6.27 + 6.08(-0.440) = 3.60.$
$\theta = \tan^{-1} 1.5 - \phi = 80° \, 04', \; \sin \theta = 0.985, \; \cos \theta = 0.173, \; \operatorname{ctn} \theta = 0.175.$
$\tan \beta = -20.07 \times 0.175 \div 3.60 = -0.978, \qquad \beta = -44° \, 02'.$

Laying off angles and scaling gives

$$a = 1.98 \text{ in.}, \qquad b = 3.05 \text{ in.}$$

$f = M(0.985 \times 3.05 \div 20.07) + M(0.173 \times 1.98 \div 3.60) = 0.243M.$
If the load is applied along the Y axis,

$$\theta = 90° - \phi = 113° \, 46',$$
$\sin \theta = 0.915, \qquad \cos \theta = -0.403, \qquad \operatorname{ctn} \theta = -0.440.$
$\tan \beta = 20.07 \times 0.440 \div 3.60 = 2.460, \qquad \beta = 67° \, 53'.$

The greatest stress is found on the inside edge of the lower leg, where

$$a = +1.18 \text{ in.}, \quad b = -3.88 \text{ in.}$$
$$f = -M(0.915 \times 3.88 \div 20.07) - M(0.403 \times 1.18 \div 3.60) = -0.309M.$$

*Examples.*—1. Find the moment of inertia of a trapezoid, bases $a$ and $b$, height $h$, about one base.

2. A 12-in. joist has two mortises cut through it, each 2 in. square, and 2 in. from edge of joist to edge of mortise. How much is that section of the joist weakened? $\frac{7}{27}$ or 26%.

3. A bridge floor is made of plates rolled to half-hexagon troughs, 6 in. face, 5.2 in. deep, 12 in. opening, $\frac{1}{2}$ in. thick. Find the resisting moment of a section 18 in. wide. $20.8f$.

4. If that floor is 14 ft. between trusses and carries two rails 5 ft. apart, each loaded with 2,000 lb. per running foot, what will be the unit stress? 7,790 lb.

5. Six thin rolled shapes, web $a$, make a hexagonal column, radius $a$, with riveted outside flanges, each $b$ in width. Prove that

$$r^2 = \frac{a^3 + 4(a+b)^3}{12(a+2b)}.$$

# CHAPTER V.

## TORSION.

**81. Torsional Moment.**—If a uniform cylindrical bar is twisted by applying equal and opposite couples or moments at two points of the axis, the planes of the couples being perpendicular to that axis, the particles on one side of a cross-section tend to rotate about the axis and past the particles on the other side of the section, thus developing a shearing stress that varies with the tendency to displacement of the particles, that is, directly as the distance of each particle from the centre. The unit shear then is constant on any *ring*, and the shearing stresses thus set up at any section make up the resisting moment to the torsional moment of the applied couple. As all cross-sections are equal and the torsional moment is constant between the two points first referred to, each longitudinal fibre will take the form of a helix.

**82. Torsional Moment of a Cylinder.**—If the unit shear at the circumference of the outer circle, Fig. 41, of radius $R_1$ and diameter $d$ is $q_1$, the value at a distance $R$ from the centre will be, by the above statement, $q = q_1 R \div R_1$. The total shearing force on the face of an infinitesimal particle whose lever-arm is $R$, and area $RdRd\theta$, will be $\dfrac{q_1}{R_1}R^2 dR d\theta$, and its moment about the centre will be $\dfrac{q_1}{R_1}R^3 dR d\theta$. Hence the resisting moment for a cylinder is

$$T = \frac{q_1}{R_1}\int_0^{R_1}\int_0^{2\pi} R^3 dR d\theta = \frac{q_1 J}{R_1} = \frac{1}{2}\pi q_1 R_1^3 = 0.196 q_1 d^3.$$

Hence the resisting moment against torsion resembles in form the resisting moment against flexure, but differs in using the polar moment of inertia of the cross-section for the rectangular one, and in having $q_1$, maximum unit shear, in place of $f$, maximum unit tension or compression.

As the rings of metal situated farthest from the centre of a shaft offer the greatest resistance to torsion, it is economical of metal to make the shaft hollow. If $d$ is the internal and $D$ the external diameter of a shaft, the resisting moment is found by inserting the expression for the polar moment of inertia of a ring in the last equation, hence

$$T = 0.196 q_1 \frac{D^4 - d^4}{D}.$$

*Example.*—Design a round shaft to transmit 500 H.P. at 100 revolutions per minute if $q_1 = 8,000$ lb. per sq. in.

$500 \times 33,000 \div 100 = 165,000$ ft.-lb. of work per revolution.

Assume this work to be performed by a force $P$ rotating about the centre of the shaft at a distance unity. The work done by $P$ in one revolution will be the force into the distance through which it acts, or $2\pi P$, and the torque, $T$, will be $P \times 1$; hence

$T = 165,000 \div 2\pi = 26,300$ ft.-lb. $= 315,000$ in.-lb.

$$d^3 = \frac{315,000}{0.196 \times 8,000} = 201, \qquad d = 6 \text{ in. nearly.}$$

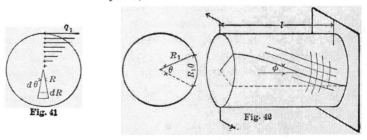

Fig. 41    Fig. 42

**83. Twist of a Cylindrical Shaft.**—If the surface of a cylindrical shaft is divided into small squares by two sets of lines, one set running along the elements and the other at right angles to them, the first set will become helices when the shaft is twisted, while the second set will remain unchanged with the result that the squares will become rhombs, as is shown in Fig. 42. The angle of distortion of the rhombs is $\phi$, which is the angle that every tangent to a helix makes with the axis of the shaft. When the free end of the shaft has rotated by the fixed end through an angle $\theta$, for small distortions such as occur in practice $\phi l = R_1 \theta$. From § 173, $\phi = q_1 \div C$, hence

$$\theta = \frac{\phi}{R_1}l = \frac{q_1}{CR_1}l = \frac{T}{CJ}l = \frac{32T}{C\pi d^4}l = 10.2\frac{T}{Cd^4}l.$$

*Example.*—A round shaft $2\frac{1}{2}$ in. diameter carries a pulley of 30 in. diameter; the difference in tension on the two parts of the belt is 1,000 lb. Then $T = 1,000 \cdot 15 = 15,000$ in.-lb. if the torsional moment is entirely carried by the section of the shaft on one side of the pulley. If the shaft is 30 ft. long and $C = 11,500,000$,

$$\theta = \frac{10.2 \cdot 15,000 \cdot 30 \cdot 12 \cdot 2^4}{11,500,000 \cdot 5^4} = 0.122.$$

To reduce this angle to degrees multiply by $180 \div \pi = 57.3$ and obtain $6° 59'$. $q_1 = 4,890$ lb. per sq. in.

**84. Non-Cylindrical Shafts.**—The formulas deduced for cylindrical shafts cannot be applied to shafts of any other form since sections which are not circular become warped under torsion and the unit shear is not proportional to its distance from the axis of the shaft. On any form of cross-section the shear at the perimeter must be tangential; for, if it is not, it can be resolved into two components, one tangential and one normal. By § 154 the normal component would necessitate an equal shear on a plane, at right angles; that is, along the surface of the shaft parallel to the axis, which is impossible. St.-Venant deduced theoretically formulas for prisms of various cross-sections under torsion. The solution is very involved and only his results will be given. In each case the maximum stress is found at points on the perimeter which lie nearest the axis. If the prism has corners or edges, the stress there is zero. The notation is the same as in § 74.

Rectangle, $T = \dfrac{5b^2h^2}{15h+9b}q_1;$

Square, $T = 0.208 q_1 h^3;$

Ellipse, $T = 0.196 q_1 b d^2;$

Equilateral triangle, $T = 0.077 q_1 h^3 = 0.050 q_1 b^3.$

84 STRUCTURAL MECHANICS.

The twist of any of these forms is very nearly

$$\theta = 40 \frac{TJ}{CS^4} l.$$

**85. Bach's Formula.**—Professor Bach deduced an expression for the torsional moment of a rectangular shaft from experiment. The sides of a rectangular steel bar were ruled into squares and the bar was twisted to give it a permanent set. The squares were distorted by the twist into rhombs and the amount of the distortion of any square was taken to be proportional to the intensity of the shear. The distortion was greatest at the middle of the longer side, at the middle of the shorter side it bore the relation of $b$ to $h$ to that of the longer side, and at the corners it was zero. The variation along one side was given very well by the ordinates to a parabola with the vertex opposite the middle of the side. The principal axes of any original right section remained straight and at right angles to each other after the bar was twisted. The unit shears at points on these axes were assumed to be proportional to their distances from the centre of the section and to act at right angles to the axes. From these assumptions this equation was deduced:

Rectangle, $T = \tfrac{2}{3} q_1 b^2 h.$   $h > b$

**86. Combined Bending and Torsion.**—A shaft is often subjected to bending in addition to torsion, and in such a case must be designed to resist both. The normal unit stress on the cross-section at the extreme fibre from the bending moment on the beam or shaft is $f$, tension on one edge, compression on the opposite edge.

The unit shear on the same cross-section at the same extreme fibre is $q_1$, and as the shears on two planes at right angles are

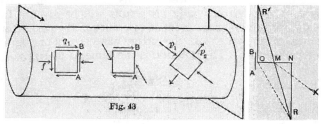

Fig. 43

equal by § 154, the stresses at the extreme fibre on a longitudinal and on a transverse plane are as shown in Fig. 43. (The plane of the bending moment is at right angles to the plane of the paper.) The principal stresses can be found by the method of § 168.

AB is the plane of cross-section; $NO = f$; $RN = q_1$. Then RO is the resultant unit stress on AB, $R'O = q_1$ on second plane revolved 90° to make the two planes and normals coincide. Draw $RR'$ connecting the extremities of the two stresses. As its middle point falls on the middle point, M, of NO,

$$p_1 = OM + MR, \quad p_2 = OM - MR. \quad MR^2 = MN^2 + NR^2.$$

$$p_1 = \tfrac{1}{2}f + \sqrt{\tfrac{1}{4}f^2 + q_1^2}, \quad \ldots \quad (1)$$

$$p_2 = \tfrac{1}{2}f - \sqrt{\tfrac{1}{4}f^2 + q_1^2}.$$

By § 152, $\quad q_{max} = \tfrac{1}{2}(p_1 - p_2) = \sqrt{\tfrac{1}{4}f^2 + q_1^2}.$

Since $M = fI \div y_1$ and $T = q_1 J \div y_1$ and since $2I = J$ for a square or a circle, (1) may be multiplied by $I \div y_1$ and transformed to

$$M_1 = \tfrac{1}{2}(M + \sqrt{M^2 + T^2}), \quad \ldots \quad (2)$$

in which $M$ = original bending moment at the section, $T$ = original torsional moment, and $M_1$ = equivalent resulting bending moment for which the shaft should be designed so that the unit stress shall not exceed $f$ when both $M$ and $T$ occur at the given section.

If (1) is multiplied by $J \div y_1$, there results

$$T_1 = M + \sqrt{M^2 + T^2} \quad \ldots \quad (3)$$

as an equivalent torsional moment in which $T_1 = p_1 J \div y_1$. Although neither equation (2) nor (3) is rational, since $p_1$ does not act on a right section, the conception of an equivalent bending moment is less objectionable than the conception of an equivalent torsional moment because $p_1$ is a normal stress. But in designing shafting it is usual to employ the same working value for unit fibre stress, $f$, as for unit shear, $q_1$; consequently the results obtained are the same whether equation (1), (2), or (3) is used.

If the deformation of the material is taken into account as in

§ 175, the new principal stresses become $p_1' = p_1 - \frac{1}{4}p_2$ and $p_2' = p_2 - \frac{1}{4}p_1$.

$$p_1' = \tfrac{3}{8}f + \tfrac{5}{4}\sqrt{\tfrac{1}{4}f^2 + q_1^2},$$
$$M_1' = \tfrac{3}{8}M + \tfrac{5}{8}\sqrt{M^2 + T^2},$$
$$T_1' = \tfrac{3}{4}M + \tfrac{5}{4}\sqrt{M^2 + T^2}.$$

*Examples.*—1. If the pulley of the previous example weighs 500 lb. and is 12 in. from the hanger, on a free end of the shaft, and the unbalanced belt-pull of 1,000 lb. is horizontal, the resultant force on the pulley is $500\sqrt{1^2 + 2^2} = 1,120$ lb., which causes a bending moment of 13,400 in.-lb. at the hanger. Then

$$M_1 = \tfrac{1}{2}\{13,400 + \sqrt{(13,400^2 + 15,000^2)}\} = 16,750 \text{ in.-lb.},$$

which will cause a fibre stress of

$$p_1 = \frac{16,750 \times 2^3}{0.098 \times 5^3} = 10,950 \text{ lb. per sq. in.}$$

2. The wooden roller of a windlass is 4 ft. between bearings. What should be its diameter to safely lift 4,000 lb. with a 2-in. rope and a crank at each end, both cranks being used and $f$ being 800 lb.?
$8\tfrac{1}{2}$ in.

3. Design a shaft to transmit 500 horse-power at 80 revolutions per min., if $q_1 = 9,000$ lb. $d = 6$ in.

4. How large a shaft will be required to resist a torsional moment of 1,600 ft.-lb. if $q_1 = 7,500$ lb.? If the shaft is 75 ft. long and $C = 11,200,000$, what will be the angle of torsion? $2\tfrac{3}{8}$ in.; $30\tfrac{1}{3}°$.

5. What torsional moment will a hollow shaft of 5 in. internal and 10 in. external diameter transmit if $q_1 = 8,000$ lb. per sq. in.? What is the size of a solid shaft to transmit the same torsional moment? What is the difference in weight between the two shafts if a bar of steel of one inch section and one foot long weighs 3.4 lb.?

# CHAPTER VI.

## FLEXURE AND DEFLECTION OF SIMPLE BEAMS.

**87. Introduction.**—As the stresses of tension and compression which make up the resisting moment at any section of a beam cause elongation and shortening of the respective longitudinal elements or layers on either side of the neutral plane, a curvature of the beam will result. This curvature will be found to depend upon the material used for the beam, upon the magnitude and distribution of the load, the span of the beam and manner of support, and upon the dimensions and form of cross-section. It is at times desirable to ascertain the amount of *deflection*, or perpendicular displacement from its original position, of any point, or of the most displaced point, of any given beam carrying a given load.

Further, the investigation of the forces and moments which act on beams supported in any other than the ways already discussed requires the use of equations that take account of the bending of the beams under these moments. There are too many unknown quantities to admit of a solution by the principles of statics alone. The required equations involve expressions for the inclination or *slope* of the tangent to the curved neutral axis of the bent beam at any point, and its *deflection*, or perpendicular displacement, at any point from its original straight line, or from a given axis. The curve assumed by the neutral axis of the bent beam is called the *elastic curve*.

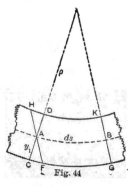

Fig. 44

**88. Formula for Curvature.**—If, through the points A and B, on the elastic curve, Fig. 44, and distant $ds$ apart, normals C D and K G to the curve of this neutral axis are drawn, the distance from A B to their intersection will be the radius of curvature $\rho$ for that portion of the curve. If through A a plane F H is passed parallel to K G, the distance F C will be the elongation, or H D will be the shortening, from the unit stress $f$, of the extreme fibre which was $ds$ long before flexure. Cross-sections plane before flexure are plane after flexure, § 62.

$A O = \rho$; $A C = y_1$; $C F = \frac{f}{E} ds$, § 10. From similar triangles A C F and O A B, $\rho : ds = y_1 : \frac{f}{E} ds$, or $\rho = \frac{E y_1}{f}$. As, by § 64, $M = \frac{fI}{y_1}$, $\frac{1}{\rho} = \frac{M}{EI}$, the reciprocal of the radius of curvature, called the *curvature* or the amount of bending at any one point.

**89. Slope and Deflection.**—If the elastic curve is referred to rectangular coordinates, $x$ being measured parallel to the original straight axis of the beam, and $v$ perpendicular to the same, the calculus gives for the radius of curvature

$$\rho = \frac{\left[1 + \left(\frac{dv}{dx}\right)^2\right]^{\frac{3}{2}}}{\frac{d^2v}{dx^2}}$$

For very slight curvature, such as exists in practical, safe beams, $\frac{dv}{dx}$ is a very small quantity, and in comparison with unity its square may be neglected. Then

$$\frac{1}{\rho} = \frac{d^2v}{dx^2} = \frac{M}{EI}.$$

As $M$ is a function of $x$, as has been seen already, the first definite integral, $\frac{dv}{dx}$, will give the tangent of the inclination or the *slope* of the tangent to the curve of the neutral axis at any point

## FLEXURE AND DEFLECTION OF SIMPLE BEAMS.

$x$, and the second integral will give $v$, the *deflection*, or perpendicular ordinate to the curve from the axis of $x$. Thus

$$\frac{dv}{dx} = \text{slope}, \quad i;$$

$$EI\frac{di}{dx} = \text{moment}, \quad M = EI\frac{d^2v}{dx^2};$$

$$\frac{dM}{dx} = \text{shear}, \quad F = EI\frac{d^3v}{dx^3};$$

$$\frac{dF}{dx} = \text{load}, \quad w = EI\frac{d^4v}{dx^4}.$$

While the following applications of the operations indicated in the last paragraph are examples only, the results in most of them will be serviceable for reference.

The student must be careful, in solving problems of this class, to use a *general value* for $M$, and *not $M$ maximum*. The origin of coordinates will be taken at a point of support, such a point being definitely located; $x$ is measured horizontally, $v$ vertically, and $-v$ denotes deflection downwards.

The greatest deflection for a given load is $v_{max}$. The greatest allowable deflection for a fibre stress $f$ is $v_1$.

If $M \div I$ is constant, the beam bends to the arc of a circle. This happens where $M$ is constant and $I$ is constant, or where $I$ varies as $M$.

*Example.*—The middle segment of a timber beam carrying a load, $W$, at each quarter-point bends to the arc of a circle. $M = \frac{1}{4}Wl;\ \rho = \frac{4EI}{Wl}$. If the beam is 6 in.$\times$12 in.$\times$20 ft., $W = 2{,}000$ lb. and $E = 1{,}500{,}000$ lb. per sq. in., then $I = \frac{1}{12} \cdot 6 \cdot 12^3 = 864$ in.$^4$ and $\rho = \frac{4 \cdot 1{,}500{,}000 \cdot 864}{2{,}000 \cdot 20 \cdot 12}$ $= 10{,}800$ in. $= 900$ ft.

**90. Beam Fixed at One End; Single Load** at the other; origin at the wall; length $= l$. Let the abscissa of any point on the elastic curve be $x$, and the ordinate $v$. At that point

$$M_x = -W(l-x).$$

Then
$$\frac{d^2v}{dx^2} = -\frac{W}{EI}(l-x).$$

The slope at the point $x$ is found by integrating.

$$i = \frac{dv}{dx} = -\frac{W}{EI}(lx - \tfrac{1}{2}x^2) + C.$$

At the point where $x=0$, the slope is zero and therefore $C=0$. Integrating again,

$$v = -\frac{W}{EI}(\tfrac{1}{2}lx^2 - \tfrac{1}{6}x^3) + C'.$$

When $x=0$, $v=0$, therefore $C'=0$. The equation just found is the equation of the elastic curve. If $x$ is made equal to $l$ in the

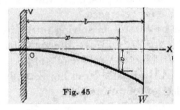

Fig. 45

above equations, the slope and the deflection at the end of the beam are found to be

$$i_{max.} = -\frac{Wl^2}{2EI};$$

$$v_{max.} = -\frac{Wl^3}{3EI}$$

To determine the maximum allowable deflection of a given beam consistent with a safe unit stress in the extreme fibre at the section of maximum bending moment, substitute, in the expression for $v_{max.}$, the value of $W$ in terms of $f$.

$$M_{max.} = -Wl = \frac{fI}{y_1}. \quad \therefore W = -\frac{fI}{y_1 l};$$

$$v_1 = -\frac{Wl^3}{3EI} = \frac{fl^2}{3Ey_1}.$$

*Example.*—If $l=60$ in., $b=4$ in., $h=8$ in., $W=800$ lb., $E=1,400,000$; $I=\dfrac{4\cdot 8^3}{12}=\dfrac{512}{3}$; max. slope $=\dfrac{800\cdot 60^2\cdot 3}{2\cdot 1,400,000\cdot 512}=0.006$; $v_{max.}=\dfrac{800\cdot 60^3\cdot 3}{3\cdot 1,400,000\cdot 512}=0.24$ in.; and if $f=1,200$ lb., maximum safe deflection $=\dfrac{1,200\times 60^2}{3\times 1,400,000\times 4}=0.26$ in.

It will be seen that, for a given weight, the maximum bending moment varies as the length $l$; the maximum slope varies as $l^2$, and the maximum deflection as $l^3$. The slope and deflection also vary inversely as $I$, or inversely as the breadth and the cube of the depth of the beam. The maximum safe deflection, however, consistent with the working unit stress $f$, varies as $l^2$, and inversely as $y_1$, or the depth of the beam. These relationships are true for other cases, as will be seen in what follows.

The ease with which problems regarding deflection are solved depends greatly upon the point taken for the origin, as it influences the value of the constants of integration.

**91. Beam Fixed at One End; Uniform Load** of $w$ per unit over the whole length, $l$; origin at the wall.

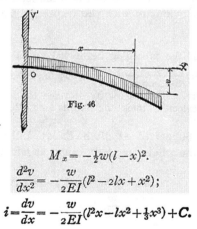

Fig. 46

$$M_x = -\tfrac{1}{2}w(l-x)^2.$$

$$\frac{d^2v}{dx^2} = -\frac{w}{2EI}(l^2 - 2lx + x^2);$$

$$i = \frac{dv}{dx} = -\frac{w}{2EI}(l^2 x - lx^2 + \tfrac{1}{3}x^3) + C.$$

When $x=0$, $i=0$; $\therefore C=0$.

$$v = -\frac{w}{2EI}(\tfrac{1}{2}l^2 x^2 - \tfrac{1}{3}lx^3 + \tfrac{1}{12}x^4) + C'.$$

When $x=0$, $v=0$; $\therefore C'=0$. When $x=l$,
$$v_{max.} = -\frac{(wl)l^3}{8EI}.$$

For maximum safe deflection consistent with unit stress, $f$, in the extreme fibre at the dangerous section,
$$M_{max.} = -(wl)\tfrac{1}{2}l = \frac{fI}{y_1}; \qquad (wl) = \frac{2fI}{y_1 l};$$
$$v_1 = -\frac{(wl)l^3}{8EI} = \frac{fl^2}{4Ey_1}.$$

**92. Combination of Uniform Load and Single Load** at one end of a beam fixed at the other end. Add the corresponding values of the two cases preceding.
$$M_{max.} = -[Wl + \tfrac{1}{2}(wl)l];$$
$$v_{max.} = -\frac{1}{EI}[\tfrac{1}{3}Wl^3 + \tfrac{1}{8}(wl)l^3] = -\frac{l^3}{3EI}(W + \tfrac{3}{8}wl).$$

Note, in the expression for $v_{max.}$, the relative deflections due to a load at the end and to the same load distributed along the beam; and compare with the respective maximum bending moments.

*Example.*—If the preceding beam weighs 50 lb., the additional deflection will be $\dfrac{50 \cdot 60^3 \cdot 3}{8 \cdot 1,400,000 \cdot 512} = 0.005$ in., too small a quantity to be of importance. In the majority of cases the weight of the beam itself may be neglected, unless the span is long.

**93. Beam Supported at Both Ends; Uniform Load** of $w$ per unit over the whole length $l$; origin at left point of support.
$$M_x = \tfrac{1}{2}w(l-x)x.$$
$$\frac{d^2v}{dx^2} = \frac{w}{2EI}(lx - x^2);$$
$$\frac{dv}{dx} = \frac{w}{2EI}(\tfrac{1}{2}lx^2 - \tfrac{1}{3}x^3 + C).$$

# FLEXURE AND DEFLECTION OF SIMPLE BEAMS. 93

From conditions of symmetry it is evident that the slope is zero when $x=\frac{1}{2}l$; hence, to determine the constant of integration, make

$$0 = \tfrac{1}{8}l^3 - \tfrac{4}{24}l^3 + C; \qquad C = -\tfrac{1}{12}l^3;$$

$$v = \frac{w}{2EI}(\tfrac{1}{6}lx^3 - \tfrac{1}{12}x^4 - \tfrac{1}{12}l^3x + C').$$

As $v=0$ when $x=0$ or $l$, $C'=0$. When $x=\tfrac{1}{2}l$,

$$v_{max.} = -\frac{5}{384}\frac{(wl)l^3}{EI}.$$

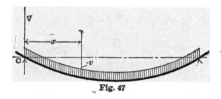

Fig. 47

For maximum safe deflection consistent with a unit stress, $f$, in the extreme fibre at the middle section,

$$\tfrac{1}{8}(wl)l = \frac{fI}{y_1}; \qquad wl = \frac{8fI}{y_1 l}; \qquad v_1 = \frac{5}{48}\frac{fl^2}{Ey_1}.$$

*Examples.*—A pine floor-joist, uniformly loaded, section $2\times 12$ in., span 14 ft.=168 in., has deflected $\tfrac{3}{4}$ in. at the middle. Is it safe? $E=1,500,000$. By the last formula,

$$\frac{3}{4} = \frac{5}{48}f\frac{14^2 \cdot 12^2}{1,500,000 \cdot 6}; \qquad f = 2,300 \text{ lb.},$$

and the beam is overloaded.

What weight is it carrying? By formula for $v_{max.}$,

$$\frac{3}{4} = \frac{5}{384}wl\frac{14^3 \cdot 12^3 \cdot 12}{1,500,000 \cdot 2 \cdot 12^3}; \qquad wl = 5,248 \text{ lb.}$$

**94. Beam Supported at Both Ends; Single Load $W$ at middle of span $l$; origin at left point of support.**

$$\frac{d^2v}{dx^2} = \frac{W}{2EI}x.$$

This expression will apply only from $x=0$ to $x=\tfrac{1}{2}l$; but, as the two halves of the deflection curve are symmetrical, the discussion of the left half will suffice.

$$i = \frac{dv}{dx} = \frac{W}{2EI}(\tfrac{1}{2}x^2 + C).$$

When $x=\tfrac{1}{2}l$, $i=0$. $\therefore C = -\tfrac{1}{8}l^2$.

$$v = \frac{W}{2EI}(\tfrac{1}{6}x^3 - \tfrac{1}{8}l^2 x + C').$$

When $x=0$, $v=0$. $\therefore C'=0$. The limit $x=l$ is not applicable.

$$v_{max.} = -\frac{Wl^3}{48EI}.$$

For maximum safe deflection consistent with a unit fibre stress, $f$,  $\quad \tfrac{1}{4}Wl = \frac{fI}{y_1};\quad W = \frac{4fI}{y_1 l};\quad v_1 = \frac{fl^2}{12Ey_1}.$

*Example.*—A 10-in. steel I beam of 33 lb. per ft. and $I=162$; span, 15 ft., $=180$ in., carries in addition a uniform load of 767 lb. per ft. of span and 6,000 lb. concentrated at the middle. What will be its deflection and the maximum unit fibre stress?

$$v_{max.} = \left(\frac{5 \cdot 800 \cdot 15}{384} + \frac{6,000}{48}\right)\frac{180^3}{29,000,000 \cdot 162} = 0.35 \text{ in.};$$

$$\frac{f \cdot 162}{5} = \left(\frac{800 \cdot 15}{8} + \frac{6,000}{4}\right)180;\quad f = 16,667 \text{ lb.}$$

## FLEXURE AND DEFLECTION OF SIMPLE BEAMS.

**95. Single Weight on Beam of Span $l$,** supported at both ends; $W$ at a given distance $a$ from the origin, which is at the left point of support.

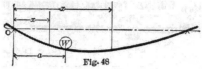

Fig. 48

As the load is eccentric, the curve of the beam is unsymmetrical, and equations must be written for each portion, $x < a$ and $x > a$.

| ON LEFT OF WEIGHT. | ON RIGHT OF WEIGHT. |
|---|---|
| $P_1 = W\dfrac{l-a}{l}$;  $M_x = \dfrac{W(l-a)}{l}x$. | $P_2 = \dfrac{Wa}{l}$;  $M_x = \dfrac{Wa}{l}(l-x)$. |
| $\dfrac{d^2v}{dx^2} = \dfrac{W}{EIl}(l-a)x$. | $\dfrac{d^2v}{dx^2} = \dfrac{W}{EIl}(al - ax)$. |
| $\dfrac{dv}{dx} = \dfrac{W}{EIl}(\tfrac{1}{2}lx^2 - \tfrac{1}{2}ax^2 + C) = i$.  (1) | $\dfrac{dv}{dx} = \dfrac{W}{EIl}(alx - \tfrac{1}{2}ax^2 + C') = i$.  (2) |
| $v = \dfrac{W}{EIl}(\tfrac{1}{6}lx^3 - \tfrac{1}{6}ax^3 + Cx + C'')$. | $v = \dfrac{W}{EIl}(\tfrac{1}{2}alx^2 - \tfrac{1}{6}ax^3 + C'x + C''')$. |
| $v = 0$, when $x = 0$; $\therefore C'' = 0$. | $v = 0$, when $x = l$; $\therefore C''' = -\tfrac{1}{3}al^3 - C'l$. |

For equations to determine the constants $C$ and $C'$, use the value $x = a$, when it will be evident that $i$ for the left segment must give the same value as does $i$ for the right segment, and $v$ at $a$ must be the same when obtained from the left column as when obtained from the right. Therefore, from (1) and (2),

$$\tfrac{1}{2}a^2l - \tfrac{1}{2}a^3 + C = a^2l - \tfrac{1}{2}a^3 + C'. \qquad C = C' + \tfrac{1}{2}a^2l.$$

$v$ at $a$ gives

$$\tfrac{1}{6}a^3l - \tfrac{1}{6}a^4 + C'a + \tfrac{1}{2}a^3l = \tfrac{1}{2}a^3l - \tfrac{1}{6}a^4 + C'a - \tfrac{1}{3}al^3 - C'l.$$
$$C' = -\tfrac{1}{6}a^3 - \tfrac{1}{3}al^2.$$
$$C''' = \tfrac{1}{6}a^3l. \qquad C = \tfrac{1}{2}a^2l - \tfrac{1}{6}a^3 - \tfrac{1}{3}al^2.$$

| | |
|---|---|
| $\dfrac{dv}{dx} = \dfrac{W}{EIl}\left(\dfrac{l-a}{2}x^2 + \dfrac{a^2l}{2} - \dfrac{a^3}{6} - \dfrac{al^2}{3}\right).$ | $\dfrac{dv}{dx} = \dfrac{W}{EIl}\left(alx - \dfrac{ax^2}{2} - \dfrac{a^3}{6} - \dfrac{al^2}{3}\right).$ |
| $v = \dfrac{W}{EIl}\left(\dfrac{l-a}{6}x^3 + \dfrac{a^2lx}{2} - \dfrac{a^3x}{6} - \dfrac{al^2x}{3}\right).$ | $v = \dfrac{W}{EIl}\left(\dfrac{alx^2}{2} - \dfrac{ax^3}{6} - \dfrac{a^3x}{6} - \dfrac{al^2x}{3} + \dfrac{a^3l}{6}\right)$ |
| | $= \dfrac{Wa}{6EIl}(l-x)[a^2 - (2l-x)x].$ |

As $a$ is assumed to be less than $\tfrac{1}{2}l$, and the substitution of $x=a$ in the value of $\dfrac{dv}{dx}$ on the left gives a slope which is negative, the point of $v_{max.}$ will be found on the right of $W$, and for that value of $x$ which makes $\dfrac{dv}{dx}$ on the right zero. Hence

$$\tfrac{1}{2}ax^2 - alx + \tfrac{1}{6}a^3 + \tfrac{1}{3}al^2 = 0.$$
$$x = l - \sqrt{\tfrac{1}{3}(l^2 - a^2)},$$

which is the distance from the left-hand support to the point of maximum deflection. Substitute in the expression for $v$ on the right, to obtain the maximum deflection.

It should be noticed that, when the weight is eccentric, the point of maximum deflection is found between the weight and the mid-span, and not at the point of maximum bending moment, which latter is under the weight.

**96. Two Equal Weights on Beam of Span $l$,** supported at the ends; each $W$, symmetrically placed, distant $a$ from one end.

This case may be solved by itself, but can be more readily treated by reference to § 95. Thus the maximum deflection will be at the middle, and can be found by making $x=\tfrac{1}{2}l$ in the above value of $v$ for the right segment and doubling the result. Then

$$v_{max.} = \frac{Wa}{3EIl} \cdot \frac{l}{2}(a^2 - \tfrac{3}{4}l^2) = -\frac{Wa}{24EI}(3l^2 - 4a^2).$$

The deflection under a weight will be given by the addition of $v$ at $a$ and $v$ at $(l-a)$ of the preceding case. Thus

$$v_a = -\frac{Wa^2}{3EIl}(l-a)^2; \quad v_{l-a} = -\frac{Wa^2}{6EIl}(l^2 - 2a^2).$$

$$v \text{ at } W = -\frac{Wa^2}{6EI}(3l - 4a).$$

## FLEXURE AND DEFLECTION OF SIMPLE BEAMS.

*Example.*—A round iron bar, 12 ft. long and 2 in. diameter, carries two weights of 200 lb. each at points 3 ft. distant from either of the two supported ends. The deflection at a weight=

$$\frac{200 \cdot 36^2 \cdot 7 \cdot 4}{6 \cdot 28{,}000{,}000 \cdot 22}(3 \cdot 12^2 - 4 \cdot 3 \cdot 12) = 0.56 \text{ in.}$$

The maximum unit bending stress is $\dfrac{200 \cdot 3 \cdot 12 \cdot 7 \cdot 4}{22} = 9{,}160$ lb.

### DEFLECTION OF BEAMS OF UNIFORM STRENGTH.

It will be apparent that a beam of uniform strength will not be so stiff as a corresponding beam of uniform section sufficient to carry safely the maximum bending moment; for the stiffness arising from the additional material in the second case is lost.

**97. Uniform Strength and Uniform Depth.** (See § 71.)—Since $M = njbh^2$ and varies as $bh^2$, and $I = n'bh^3$ and varies as $bh^3$, $M \div I$ varies as $1 \div h$. But if $h$ is constant, $M \div I$ is constant and $\dfrac{d^2v}{dx^2}$ is constant. Therefore all beams of this class bend to the arc of a circle.

I. Beam fixed at one end only, and loaded with $W$ at the other. Fig. 23.

If $I_0$ is the moment of inertia at the largest section, which is in this case at the wall, and $I$ is the variable moment of inertia,

$$\frac{I}{I_0} = \frac{b}{b_0} = \frac{l-x}{l}; \qquad \frac{l-x}{I} = \frac{l}{I_0}.$$

$$\frac{d^2v}{dx^2} = \frac{M}{EI} = -\frac{W}{EI}(l-x) = -\frac{Wl}{EI_0}.$$

$$\frac{dv}{dx} = -\frac{Wl}{EI_0}x + C; \qquad C = 0.$$

$$v = -\frac{Wl}{2EI_0}x^2 + C'; \qquad C' = 0.$$

$$v_{max.} = -\frac{Wl^3}{2EI_0},$$

a deflection 50 per cent. in excess of that of the corresponding uniform beam, while the maximum slope is twice as great.

*Examples.*—If a triangular sheet of metal, like the dotted triangle in Fig. 49, is cut into strips, as represented by the dotted lines, and

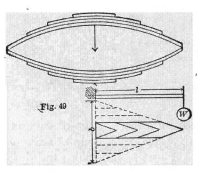

Fig. 49

these strips are superimposed as shown above, the strips, if fixed at the ends, and subjected to $W$ as shown, will tend to bend in arcs of circles, and will remain approximately in contact. If $l=10$ in., $b=4$ in., $h=\frac{1}{4}$ in., $W=400$ lb. and $E=28{,}000{,}000$, the deflection will be $\dfrac{400 \cdot 10^3 \cdot 4^3 \cdot 12}{2 \cdot 28{,}000{,}000 \cdot 4} = 1.37$ in.

An elliptical steel spring 2 ft. long, of 4 layers as shown, each 2 in. broad and $\frac{1}{8}$ in. thick, under a load of 100 lb. at its middle, will, if $E=29{,}000{,}000$, deflect $\dfrac{100 \cdot 12^3 \cdot 8^3 \cdot 12}{2 \cdot 29{,}000{,}000 \cdot 8} = 2.3$ in. The maximum unit fibre stress will be $\dfrac{50 \cdot 12 \cdot 6 \cdot 8^2}{8} = 28{,}800$ lb. Note that one-half of the weight is found at each hinge, and that the deflection of one arm is doubled by the use of two springs as shown.

II. Beam fixed at one end only and uniformly loaded with $w$ per unit. Fig. 25.

$$\frac{I}{I_0} = \frac{b}{b_0} = \frac{(l-x)^2}{l^2}. \qquad \frac{(l-x)^2}{I} = \frac{l^2}{I_0}.$$

$$\frac{d^2v}{dx^2} = -\frac{w}{2EI}(l-x)^2 = -\frac{wl^2}{2EI_0};$$

$$\frac{dv}{dx} = -\frac{wl^2}{2EI_0}x + C; \qquad C=0.$$

$$v = -\frac{wl^2}{4EI_0}x^2 + C'; \qquad C'=0.$$

$$v_{max.} = -\frac{wl^4}{4EI_0},$$

a deflection twice that of the corresponding uniform beam.

## FLEXURE AND DEFLECTION OF SIMPLE BEAMS.

In these two cases there are no constants of integration, since $\frac{dv}{dx}$ and $v=0$ when $x=0$.

III. Beam supported at both ends and carrying $W$ at middle. Fig. 27.

$$I:I_0 = b:b_0 = x:\tfrac{1}{2}l.$$

$$\frac{d^2v}{dx^2} = \frac{Wx}{2EI} = \frac{Wl}{4EI_0};$$

$$\frac{dv}{dx} = \frac{Wl}{4EI_0}(x+C).$$

Slope $=0$ when $x=\tfrac{1}{2}l$; $\therefore C = -\tfrac{1}{2}l.$

$$v = \frac{Wl}{4EI_0}(\tfrac{1}{2}x^2 - \tfrac{1}{2}lx) + C'; \qquad C' = 0.$$

$$v_{max.} = -\frac{Wl^3}{32EI_0},$$

a deflection 50 per cent. greater than for a corresponding uniform beam.

IV. Beam supported at both ends, and uniformly loaded with $w$ per unit of length. Fig. 29.

$$I:I_0 = b:b_0 = x(l-x):\tfrac{1}{4}l^2.$$

$$\frac{d^2v}{dx^2} = \frac{w}{2EI}(lx-x^2) = \frac{wl^2}{8EI_0};$$

$$\frac{dv}{dx} = \frac{wl^2}{8EI_0}(x+C); \qquad C = -\tfrac{1}{2}l.$$

$$v = \frac{wl^2}{8EI_0}(\tfrac{1}{2}x^2 - \tfrac{1}{2}lx + C'); \qquad C' = 0.$$

$$v_{max.} = -\frac{wl^4}{64EI_0},$$

a deflection 20 per cent. greater than for a corresponding uniform beam.

**98. Uniform Strength and Uniform Breadth.**—In these cases, as $b$ is constant, $I:I_0 = h^3:h_0^3$ or $I = I_0 \frac{h^3}{h_0^3}$.

V. Beam fixed at one end only, and loaded with $W$ at the other. Fig. 24.

$$\frac{h^2}{h_0^2}=\frac{l-x}{l}. \qquad I=I_0\frac{h^3}{h_0^3}=I_0\frac{(l-x)^{\frac{3}{2}}}{l^{\frac{3}{2}}}.$$

$$\frac{d^2v}{dx^2}=-\frac{W}{EI}(l-x)=-\frac{Wl^{\frac{3}{2}}}{EI_0}(l-x)^{-\frac{1}{2}};$$

$$\frac{dv}{dx}=-\frac{Wl^{\frac{3}{2}}}{EI_0}(-2(l-x)^{\frac{1}{2}}+C); \qquad C=2l^{\frac{1}{2}}.$$

$$v_{max.}=\frac{2Wl^{\frac{3}{2}}}{EI_0}\int_0^l[(l-x)^{\frac{1}{2}}-l^{\frac{1}{2}}]dx$$

$$=\frac{2Wl^{\frac{3}{2}}}{EI_0}[-\tfrac{2}{3}(l-l)^{\frac{3}{2}}-l^{\frac{3}{2}}+\tfrac{2}{3}l^{\frac{3}{2}}]=-\frac{2}{3}\frac{Wl^3}{EI_0},$$

or twice the deflection of a corresponding uniform beam.

VI. Beam fixed at one end only and uniformly loaded with $w$ per unit of length. Fig. 26.

$$\frac{h}{h_0}=\frac{l-x}{l}. \qquad I=I_0\frac{(l-x)^3}{l^3}.$$

$$\frac{d^2v}{dx^2}=-\frac{w}{2EI}(l-x)^2=-\frac{wl^3}{2EI_0}(l-x)^{-1};$$

$$\frac{dv}{dx}=\frac{wl^3}{2EI_0}[\log(l-x)-\log l].$$

$$v_{max.}=\frac{wl^3}{2EI_0}\int_0^l[\log(l-x)-\log l]dx*$$

$$=\frac{wl^3}{2EI_0}\Big[x\log(l-x)-x-l\log(l-x)-x\log l\Big]_0^l$$

$$=\frac{wl^3}{2EI_0}(l\log 0-l-l\log 0-l\log l+l\log l)=-\frac{wl^4}{2EI_0},$$

or four times the deflection of a corresponding uniform beam.

---

* Log $(l-x)=u$; $dx=dv$; $x=v$; $du=-\frac{dx}{l-x}$.

$\therefore \int\log(l-x)dx=x\log(l-x)+\int\frac{x}{l-x}dx$. By division, $\frac{x}{l-x}=-1+\frac{l}{l-x}$;

$\therefore \int\frac{x}{l-x}dx=-\int dx+l\int\frac{dx}{l-x}=-x-l\log(l-x).$

# FLEXURE AND DEFLECTION OF SIMPLE BEAMS.

**VII.** Beam supported at both ends and carrying $W$ at middle. Fig. 28.

$$\frac{h^2}{h_0^2} = \frac{x}{\frac{1}{2}l}. \qquad I = I_0\sqrt{\frac{8x^3}{l^3}}.$$

$$\frac{d^2v}{dx^2} = \frac{Wx}{2EI} = \frac{Wl^{\frac{3}{2}}}{4\sqrt{2}EI_0}x^{-\frac{1}{2}};$$

$$\frac{dv}{dx} = \frac{Wl^{\frac{3}{2}}}{4\sqrt{2}EI_0}(2x^{\frac{1}{2}}+C).$$

When $x = \frac{1}{2}l$, $\frac{dv}{dx} = 0$, and $C = -2\sqrt{\frac{1}{2}l}$.

$$v_{max.} = \frac{Wl^{\frac{3}{2}}}{2\sqrt{2}EI_0}\int_0^{\frac{l}{2}}(x^{\frac{1}{2}} - \sqrt{\frac{1}{2}l})dx$$

$$= \frac{Wl^{\frac{3}{2}}}{2\sqrt{2}EI_0}\left[\frac{2}{3}x^{\frac{3}{2}} - \sqrt{\frac{1}{2}l}\,x\right]_0^{\frac{l}{2}}$$

$$= \frac{Wl^{\frac{3}{2}}}{2\sqrt{2}EI_0}\left(\frac{2}{3}\frac{l^{\frac{3}{2}}}{2\sqrt{2}} - \frac{l^{\frac{3}{2}}}{2\sqrt{2}}\right) = -\frac{Wl^3}{24EI_0},$$

or twice the deflection of a corresponding uniform beam.

**VIII.** Beam supported at both ends, and uniformly loaded with $w$ per unit of length. Fig. 30.

$$\frac{h^2}{h_0^2} = \frac{x(l-x)}{\frac{1}{4}l^2}. \qquad I = I_0\frac{8\sqrt{x^3(l-x)^3}}{l^3}.$$

$$\frac{d^2v}{dx^2} = \frac{w}{2EI}(lx - x^2) = \frac{wl^3}{16EI_0}(lx - x^2)^{-\frac{1}{2}};$$

$$\frac{dv}{dx} = \frac{wl^3}{16EI_0}\left(\text{versin}^{-1}\frac{2x}{l} + C\right);$$

$$\frac{dv}{dx} = 0, \text{ when } x = \frac{1}{2}l; \quad \therefore C = -\text{versin}^{-1} 1 = -\frac{1}{2}\pi.$$

$$v_{max.} = \frac{wl^3}{16EI_0}\int_0^{\frac{1}{2}l}\left(\text{versin}^{-1}\frac{2x}{l} - \frac{1}{2}\pi\right)dx$$

$$= \frac{wl^3}{16EI_0}\left[(x - \frac{1}{2}l)\text{versin}^{-1}\frac{2x}{l} + \sqrt{(lx - x^2)} - \frac{1}{2}\pi x\right]_0^{\frac{1}{2}l}$$

$$= \frac{wl^3}{16EI_0}\left[\frac{l}{2} - \frac{\pi}{4}l\right] = -\frac{0.2854wl^4}{16EI_0} = -\frac{0.018wl^4}{EI_0},$$

or 37 per cent. greater deflection than for a corresponding uniform beam.

Other beams might be analyzed where both $b$ and $h$ varied at the same time. The method of analysis would agree with the above; but the cases are not of sufficient practical value to warrant their discussion here.

**99. Sandwich Beam.**—If a beam is made up from two materials placed side by side, as when a plate of steel is bolted securely between two sticks of timber, the distribution of the load between the several pieces can be found from the consideration that they are compelled to deflect equally. If the pieces are of the same depth, the extreme fibres of the two materials must undergo the same change of length; or by § 10, if the subscripts $w$ and $s$ stand for wood and steel respectively,

$$\lambda_w = \lambda_s = \frac{f_w}{E_w} = \frac{f_s}{E_s}. \qquad \frac{f_w}{f_s} = \frac{E_w}{E_s}.$$

The resisting moment of such a beam is

$$M = \tfrac{1}{6} f_w b_w h^2 + \tfrac{1}{6} f_s b_s h^2 = \tfrac{1}{6} f_w \left( b_w + \frac{E_s}{E_w} b_s \right) h^2.$$

Such beams are easily figured by substituting for the steel an equivalent breadth of wood and proceeding as if the beam were entirely of wood or vice versa.

If $E$ for timber is 1,400,000 and for steel 28,000,000, the ratio of the stresses will be $\tfrac{1}{20}$; and if $f_w = 800$ lb., $f_s = 16,000$ lb. on the square inch.

*Example.*—Two 4×10-in. sticks of timber with a 10×¼-in. steel plate firmly bolted between them will have a value of

$$M = \frac{(800 \cdot 8 + 16{,}000 \cdot \tfrac{1}{4}) 10^2}{6} = 173{,}333 \text{ in.-lb.,}$$

the plate supplying $\tfrac{5}{13}$ of the amount. The combination for a span of 10 ft. would safely carry $\dfrac{4M}{120} = 5{,}778$ lb. load at center, or 11,555 lb. distributed load, in place of 3,555 or 7,110 lb. for the timber alone.

## FLEXURE AND DEFLECTION OF SIMPLE BEAMS.   103

**100. Resilience of a Beam.**—If a beam carries a single weight $W$, and the deflection under that weight is $v_1$, the external work done by that static load on the beam is $\tfrac{1}{2}Wv_1$. If this value of $v_1$ is that which causes the maximum safe unit stress $f$, the quantity $\tfrac{1}{2}Wv_1$ is known as the *resilience* of the beam, or the *energy of the greatest shock* which the beam can bear without injury, being the product of a weight into the height from which it must fall to produce the shock in question. For a beam supported at both ends, loaded in the middle, and of rectangular section,

$$v_1 = \frac{fl^2}{6Eh}, \quad \text{and} \quad W = \frac{2fbh^2}{3l}.$$

$$\tfrac{1}{2}Wv_1 = \frac{1}{2} \cdot \frac{2}{3} \cdot \frac{fbh^2}{l} \cdot \frac{fl^2}{6Eh} = \frac{1}{18} \cdot \frac{f^2}{E} \cdot bhl.$$

The allowable shock, or the resilience, is therefore proportioned to $f^2 \div E$, which is known as the *modulus of resilience* of the material, and to the *volume* of the beam. These relationships hold for other sections, and beams loaded and supported differently. The above formula should not be rigorously applied to a drop test, unless $f$ is below the yield-point.

*Example.*—A $2 \times 2$ in.-bar of steel, 5 ft. between supports, if $f = 16{,}000$ lb., ought not to be subjected, from a central weight, to more than $\dfrac{1}{18} \dfrac{16{,}000^2 \cdot 2^2 \cdot 60}{29{,}000{,}000} = 118$ in.-lb. of energy.

If the load is distributed over a similar beam, the deflection at each point will be $v$, and the total work done will be $\tfrac{1}{2}\int vw\,dx$. If $w$ is uniform, and the beam is supported at its ends,

$$\tfrac{1}{2}w\int_0^l v\,dx = \frac{w^2}{4EI}\int_0^l \left(\frac{lx^3}{6} - \frac{x^4}{12} - \frac{l^3 x}{12}\right)dx = \frac{w^2 l^5}{240EI}.$$

As 
$$\frac{wl^2}{8} = \frac{fI}{y_1} \quad \text{and} \quad v_1 = \frac{5}{48}\frac{fl^2}{Ey_1},$$

the above expression becomes

$$\text{Resilience} = \frac{w^2 l^5}{240EI} = \frac{8}{25}(wl)v_1.$$

*Example.*—A weighted wheel of 1,000 lb. drops $\frac{1}{2}$ in. by reason of a pebble in its path at the middle of a beam $3 \times 12$ in., 15 ft. span. If $E = 1,400,000$ to find $f$:

$$\text{External work} = 1,000(\tfrac{1}{2}+v) = \frac{3 \cdot 12 \cdot 180}{18 \cdot E} f^2 = \frac{18}{70,000} f^2.$$

$$v = \frac{1}{12}\frac{fl^2}{Ey_1} = \frac{180 \cdot 180}{6 \cdot 12 \cdot E} f = \frac{9}{28,000} f.$$

$$500 + \frac{9}{28} f = \frac{18}{70,000} f^2. \qquad f = 2,150 \text{ lb. per sq. in.}$$

Resulting deflection $= 0.69$ in. Static unit stress would be 625 lb. and $v = 0.2$ in. In an actual bridge the shock is distributed more or less in the floor and adjacent beams.

**101. Internal Work.**—The internal work done in a beam may be divided into two parts, that of extending and compressing the fibres, and that of distorting them, the first being due to the bending stresses, and the second to the shearing stresses. The work of bending will be found first.

Let the cross-section be constant. The unit stress at any point of a cross-section $= p$; the force on a layer $zdy = pzdy$. The elongation or shortening of a fibre unity of section and $dx$ long by the unit stress $p = pdx \div E$. The work done in stretching or shortening the volume $zdydx = \frac{1}{2} \cdot \frac{p^2}{E} \cdot zdydx$. But $p = \frac{f}{y_1} y = \frac{M}{I} y$. The work done on so much of the beam as is included between two cross-sections $dx$ apart will be

$$\frac{1}{2} \cdot \frac{M^2}{EI^2} dx \int_{-y_1}^{+y_1} y^2 z dy = \frac{M^2}{2EI} dx.$$

Substitute the value of $M$ for a particular case in terms of $x$ and integrate for the whole length of the beam. Thus for a beam supported at ends and loaded with $W$ at the middle, $M = \frac{1}{2}Wx$ at any point distant $x$ from one end for values of $x$ between $0$ and $\frac{1}{2}l$. Then

$$\frac{2}{2EI} \int_0^{\frac{1}{2}l} M^2 dx = \frac{W^2}{4EI} \int_0^{\frac{1}{2}l} x^2 dx = \frac{W^2 l^3}{96 EI}.$$

# FLEXURE AND DEFLECTION OF SIMPLE BEAMS. 105

The internal work due to shear is not readily found unless a simple form of cross-section is chosen. If the section is rectangular the shear at any point on the cross-section is, by § 72,

$$q = \frac{6F}{bh^3}(\tfrac{1}{4}h^2 - y^2).$$

The distortion in a length $dx$ is $qdx \div C$ and the work done upon a volume $zdydx$ is $\frac{q}{2} \cdot \frac{qdx}{C} \cdot zdy$. The work done upon the volumes between two sections $dx$ apart is

$$\frac{dx}{2C}\int_{-\tfrac{1}{2}h}^{+\tfrac{1}{2}h} q^2 z dy = \frac{18F^2}{bh^6 C}dx \int_{-\tfrac{1}{2}h}^{+\tfrac{1}{2}h}(\tfrac{1}{4}h^2 - y^2)^2 dy = \frac{3}{5}\frac{F^2}{bhC}dx.$$

For a beam carrying a single load at the centre, as above, $F = \tfrac{1}{2}W$, a constant, and the work done by the shearing forces is

$$\frac{3}{20}\frac{W^2 l}{bhC}.$$

**102. Deflection Due to Shear.**—In the cases of flexure so far treated the bending moment only has been taken into account in finding the deflection. The shearing stresses, however, cause an additional deflection which is generally too small to be of practical account. The deflection of a rectangular beam carrying a single load at the centre can be found from the results of the preceding section since the internal work must be equal to the external.

$$\tfrac{1}{2}Wv = \frac{W^2 l^3}{96EI} + \frac{3W^2 l}{20bhC};$$

$$v = \frac{Wl^3}{48EI} + \frac{3Wl}{10bhC}.$$

The first of these terms is the deflection when the shear is not taken into account, and it has the same value as was obtained in § 94. The ratio of the second term to the first is $\frac{6Eh^2}{5Cl^2}$, which shows that in rectangular beams of ordinary proportions the shearing deflection is but a small proportion of the whole.

*Examples.*—1. What is the deflection at the middle of a $2\times 12$-in. pine joist of 12 ft.$=144$ in. span, supported at ends and uniformly loaded with 3,200 lb.? $E=1,600,000$.     0.27 in.

2. What is the deflection if the load is at the middle?   0.432 in.

3. Find the stiffest rectangular cross-section, $bh$, to be obtained from a round log of diameter $d$.     $b=\tfrac{1}{2}d$.

4. A $4\times 6$-in. joist laid flatwise on supports 10 ft. apart is loaded with 1,000 lb. at the middle. The deflection is found to be 0.7 in. What is $E$?     1,607,000.

5. What is the maximum safe deflection of a 12-in. floor-joist, 14 ft. span, if $f=1,200$ lb. and $E=1,600,000$?

    0.37 in. for uniform load; 0.29 in. for load at middle.

6. What is the diameter of the smallest circle into which $\tfrac{1}{4}$-in. steel wire can be coiled without exceeding a fibre stress of 20,000 lb. per sq. in.?

## CHAPTER VII.

### RESTRAINED AND CONTINUOUS BEAMS.

**103. Restrained Beams.**—When a beam is kept from *rotating* at one or both points of support, by being built into a wall, or by the application of a moment of such a magnitude that the tangent to the curve of the neutral plane at the point of support is forced to remain in its original direction (commonly horizontal) at such point, the beam is termed *fixed* at one or both supports. The magnitude of the moment at the point of support depends upon the span, the load and its position. It is the existence of this at present unknown moment which calls for the application of deflection equations to the solution of such problems as those which follow, there being too many unknown quantities to permit the treatment of Chapter III.

In applying the results obtained in the following cases to actual problems, one should feel sure that the beam is definitely fixed in direction at the given point. Otherwise the values of $M$, $F$, and $v$ will be only approximately true.

**104. Beam of Span $l$, Carrying a Single Weight $W$** in the middle and supported and fixed at both ends. Origin at left support. Fig. 50.

The reactions and end moments are now unknown. The beam may be considered either as built in at its ends (as at the right in above figure), or as having an unknown couple or moment $Qb$ applied at each point of support (as at the left), of a magnitude just sufficient to keep the tangent there horizontal.

The reaction at either end will then be $\frac{1}{2}W+Q$, while the shear between the points of support will still be $\pm\frac{1}{2}W$. For values of $x < \frac{1}{2}l$,

$$M_x = -Q(b+x) + (\tfrac{1}{2}W+Q)x = \tfrac{1}{2}Wx - Qb.$$

If this value is compared with that of $M_x$ in § 94, it is seen that a *constant* subtractive or negative moment is now felt over the whole span in combination with the usual $\frac{1}{2}Wx$.

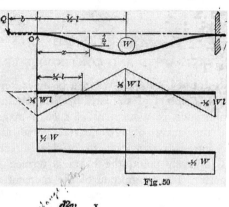

Fig. 50

$$\frac{d^2v}{dx^2} = \frac{1}{EI}(\tfrac{1}{2}Wx - Qb).$$

$$\frac{dv}{dx} = \frac{1}{EI}(\tfrac{1}{4}Wx^2 - Qbx + C).$$

Slope is zero when $x=0$; $\therefore C=0$. Also slope is zero when $x=\tfrac{1}{2}l$.

$$0 = \tfrac{1}{16}Wl^2 - \tfrac{1}{2}Qbl. \qquad -Qb = -\tfrac{1}{8}Wl,$$

the negative bending moment at either end. That it is negative appears by making $x=0$ in $M_x$ above.

If this value of $Qb$ is substituted in the first equation, giving $M_x = \tfrac{1}{2}W(x - \tfrac{1}{4}l)$, the point of contraflexure is located at $x = \tfrac{1}{4}l$, and the bending moment at middle, where $x = \tfrac{1}{2}l$, is $M_{max.} = +\tfrac{1}{8}Wl$, or one-half the amount in § 94. Substituting the value of $Qb$ in the equation for slope,

$$\frac{dv}{dx} = \frac{1}{EI}(\tfrac{1}{4}Wx^2 - \tfrac{1}{8}Wlx) = \frac{W}{4EI}(x^2 - \tfrac{1}{2}lx).$$

$$v = \frac{W}{4EI}(\tfrac{1}{3}x^3 - \tfrac{1}{4}lx^2 + C').$$

As $v = 0$ when $x = 0$, $C' = 0$. When $x = \tfrac{1}{2}l$,

## RESTRAINED AND CONTINUOUS BEAMS.

$$v_{max.} = \frac{W}{4EI}\left(\frac{l^3}{24} - \frac{l^3}{16}\right) = -\frac{Wl^3}{192EI}.$$

The beam is therefore *four* times as stiff as when only supported at ends.

As $\dfrac{fI}{y_1} = \dfrac{Wl}{8}$, $W = \dfrac{8fI}{y_1 l}$ and $v_1 = \dfrac{fl^2}{24Ey_1}$,

so that only one-half the deflection is allowable that is permitted in § 94, but the beam may safely carry twice the load.

It is useful to notice that this beam has a bending moment at the middle equal to that which would exist there if the beam were cut at the points of contraflexure and simply supported at those points, and that the two end segments, of length $\frac{1}{4}l$, act like two cantilevers each carrying $\frac{1}{2}W$, the shear at the point of contraflexure.

If the weight were not at the middle the moments at the two ends would differ, equations would be needed for each of the two segments, and the solution, while possible, would be much more complicated.

*Example.*—A wooden beam, 6 in. square and $7\frac{1}{2}$ ft. span, is built into the wall at both ends. A central weight of 3,000 lb. will give a max. fibre stress of $\dfrac{3{,}000 \cdot 90 \cdot 6}{8 \cdot 6^3} = 937\frac{1}{2}$ lb. per sq. in. at the middle and both ends. The deflection will be $\dfrac{3{,}000 \cdot 90^3 \cdot 12}{192 \cdot 1{,}500{,}000 \cdot 6^4} = 0.07$ in. if $E = 1{,}500{,}000$. The allowable deflection for $f = 1{,}200$ is $\dfrac{1{,}200 \cdot 90 \cdot 90}{24 \cdot 1{,}500{,}000 \cdot 3} = 0.09$ in., and max. allowable $W = 3{,}000 \cdot \frac{9}{7} = 3{,}860$ lb.

**105. Beam of Span $l$, Uniform Load** of $w$ per unit over the whole span, fixed at both ends. Origin at left support. Fig. 51.

As in the previous case, the reaction at either end may be represented by $\frac{1}{2}wl + Q$. The shear at $x$ is $\frac{1}{2}wl - wx$, which expression changes sign at the middle and at either point of support; hence at those places will be found $M_{max}$.

$$M_x = (\tfrac{1}{2}wl + Q)x - \tfrac{1}{2}wx^2 - Q(b + x) = \tfrac{1}{2}wlx - \tfrac{1}{2}wx^2 - Qb.$$

Compare with § 93.

$$\frac{d^2v}{dx^2} = \frac{1}{EI}(\tfrac{1}{2}wlx - \tfrac{1}{2}wx^2 - Qb);$$

$$\frac{dv}{dx} = \frac{1}{EI}(\tfrac{1}{4}wlx^2 - \tfrac{1}{6}wx^3 - Qbx + C).$$

$\frac{dv}{dx} = 0$ when $x = 0$; $\therefore C = 0$. $\quad \frac{dv}{dx} = 0$ when $x = l$;

$$\therefore \frac{wl^3}{4} - \frac{wl^3}{6} - Qbl = 0. \quad -Qb = -\frac{wl^2}{12},$$

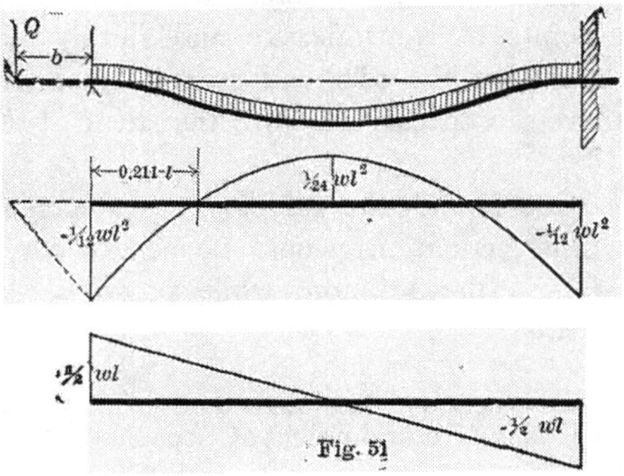

Fig. 51.

the negative moment at each point of support. If $x = \tfrac{1}{2}l$,

$$M \text{ at middle} = wl^2(\tfrac{1}{4} - \tfrac{1}{8} - \tfrac{1}{12}) = \frac{wl^2}{24}.$$

Substitute the value of $Qb$ and get

$$M_x = -\tfrac{1}{2}w(x^2 - lx + \tfrac{1}{6}l^2);$$

$$\frac{dv}{dx} = -\frac{w}{2EI}(\tfrac{1}{3}x^3 - \tfrac{1}{2}lx^2 + \tfrac{1}{6}l^2x);$$

$$v = -\frac{w}{2EI}\left(\frac{x^4}{12} - \frac{lx^3}{6} + \frac{l^2x^2}{12} + C'\right).$$

Since $v=0$ when $x=0$, $C'=0$. When $x=\frac{1}{2}l$,

$$v_{max.} = -\frac{w}{2EI}\left(\frac{l^4}{192} - \frac{l^4}{48} + \frac{l^4}{48}\right) = -\frac{(wl)l^3}{384EI},$$

which is one-fifth the value of § 93.

The points of contraflexure occur where $M_x = 0$;

$$\therefore x^2 - lx + \tfrac{1}{6}l^2 = 0; \qquad x = \tfrac{1}{2}l \pm \tfrac{1}{2}l\sqrt{\tfrac{1}{3}}.$$

The second term is the distance from the middle, each way, to the points of contraflexure. If $M$ is calculated for the middle point of a span $l \div \sqrt{3}$, it will prove to be $wl^2 \div 24$, as above. Since

$$\frac{(wl)l}{12} = \frac{fI}{y_1}, \quad wl = \frac{12fI}{y_1 l}, \quad \text{and} \quad v_1 = \frac{fl^2}{32Ey_1},$$

or 0.3 as much as in § 93. The beam may safely carry, however, 50 per cent. more load.

*Example.*—An 8-in. I beam of 12 ft. span, carrying 1,000 lb. per foot, if firmly fixed at both ends and not to have a larger unit stress than 12,000 lb., should have a value of $I = \dfrac{1{,}000 \cdot 12 \cdot 12 \cdot 12 \cdot 4}{12 \cdot 12{,}000} = 48$. An 8-in. steel beam, 18 lb. to the foot, $I = 57.8$, will satisfy the requirement, the load then being 1,018 lb. per foot. The deflection will be 0.06 in.

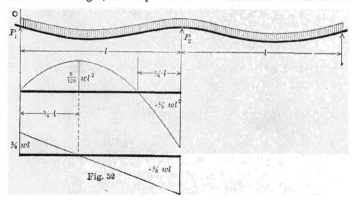

Fig. 52

**106. Beam of Span $l$,** fixed or horizontal at $P_2$, supported at $P_1$ and carrying a uniform load of $w$ per unit. Fig. 52. Origin at $P_1$ and reaction unknown.

## STRUCTURAL MECHANICS.

It will be seen from the sketch that a beam of length $2l$, resting upon three equidistant supports in the same straight line, will come under this case.

$$\frac{d^2v}{dx^2} = \frac{1}{EI}(P_1 x - \tfrac{1}{2}wx^2);$$

$$\frac{dv}{dx} = \frac{1}{EI}(\tfrac{1}{2}P_1 x^2 - \tfrac{1}{6}wx^3 + C).$$

$\frac{dv}{dx} = 0$ when $x = l$; $\therefore C = \tfrac{1}{6}wl^3 - \tfrac{1}{2}P_1 l^2$.

$$v = \frac{1}{EI}\left(\frac{P_1 x^3}{6} - \frac{wx^4}{24} + \frac{wl^3 x}{6} - \frac{P_1 l^2 x}{2} + C'\right).$$

$v = 0$ when $x = 0$; $\therefore C' = 0$. If $x = l$, $v = 0$, and
$P_1 = (\tfrac{1}{24} - \tfrac{1}{6})wl^4 \div (\tfrac{1}{6} - \tfrac{1}{2})l^3 = \tfrac{3}{8}wl$.  $F_x = \tfrac{3}{8}wl - wx$.

Substitute this value of $P_1$ in the above equations.

$$\frac{dv}{dx} = \frac{w}{2EI}(\tfrac{3}{8}lx^2 - \tfrac{1}{3}x^3 - \tfrac{1}{24}l^3).$$

$$v = \frac{w}{8EI}(\tfrac{1}{2}lx^3 - \tfrac{1}{3}x^4 - \tfrac{1}{6}l^3 x).$$

For $v_{max.}$ make $\frac{dv}{dx} = 0$, or $\tfrac{3}{8}lx^2 - \tfrac{1}{3}x^3 - \tfrac{1}{24}l^3 = 0$.

$$8x^3 - 9lx^2 + l^3 = 0.$$

As a minimum value of $v$, or $v = 0$, occurs for $x = l$, divide by $x - l$ and obtain

$$8x^2 - lx - l^2 = 0, \quad \text{or} \quad x = l\frac{1 \pm \sqrt{33}}{16} = 0.4215l.$$

Then $$v_{max.} = -0.0054\frac{(wl)l^3}{EI}.$$

To find points of $M_{max.}$ put $F_x = 0$, or $x = \tfrac{3}{8}l$.
Also, by inspection, $M_{max.}$ when $x = l$.
For $x = \tfrac{3}{8}l$, $M_{max.} = (\tfrac{9}{64} - \tfrac{9}{128})wl^2 = \tfrac{9}{128}wl^2$.

For $x=l$, $M_{max.} = (\frac{3}{8}-\frac{1}{2})wl^2 = -\frac{1}{8}wl^2$.

For the point of contraflexure $\frac{3}{8}lx - \frac{1}{2}x^2 = 0$, or $x = \frac{3}{4}l$, as was to be expected from the position of the point of maximum positive $M$.

Note again that the point of maximum bending moment is *not* the point of maximum deflection.

It will be seen that a continuous beam of two equal spans $l$, uniformly loaded with $w$ per unit, has end reactions of $\frac{3}{8}wl$ and a central reaction of $2 \times \frac{5}{8}wl = \frac{5}{4}wl$; that points of contraflexure divide each span at $\frac{1}{4}l$ from the middle pier, and that the bending moment at the middle of the remaining segment of $\frac{3}{4}l$ is, as above, $\frac{3}{4} \cdot \frac{3}{4} \cdot \frac{1}{8}wl^2 = \frac{9}{128}wl^2$. It will also be seen that, since the bending moment at $P_2$ is $-\frac{1}{8}wl^2$, a uniform beam, continuous over two equal spans, each $l$, is no stronger than the same beam of span $l$ with the same uniform load. It is, however, about two and a half times as stiff.

*Example.*—A girder spanning two equal openings of 15 ft. and carrying a 16-in. brick wall 10 ft. high of 110 lb. per cubic ft. will throw a load of $\dfrac{5 \cdot 4 \cdot 110 \cdot 10 \cdot 15}{4 \cdot 3} = 27{,}500$ lb. on the middle post and must resist a bending moment of $\dfrac{4 \cdot 110 \cdot 10 \cdot 15 \cdot 15}{3 \cdot 8} = 41{,}250$ ft.-lb.

**107. Two-span Beam, with Middle Support Lowered.**—A uniform beam, uniformly loaded and supported at its ends, will have a certain deflection at the middle which can be calculated. If the middle point is then lifted by a jack until returned to the straight line through the two end supports, the pressure on the jack, by § 106, will be five-eighths of the load on the beam. Since deflection is proportional to the weight, other things being equal, if the jack is then lowered one-fifth of the first deflection referred to, the pressure on the jack will be reduced one-fifth, or to one-half of the load on the beam. Hence, if a uniformly loaded beam of two equal, continuous spans has its middle support lower than those at its ends by one-fifth of the above deflection, the middle reaction will be one-half the whole weight, the *bending moment* will be *zero* at the middle and the beam may be cut at that point without disturbance of the forces.

**108. Beam of Span $l$,** fixed at left and supported at right end, and carrying a single weight $W$ at a distance $a$ from the fixed end. Fig. 53. Origin at fixed end.

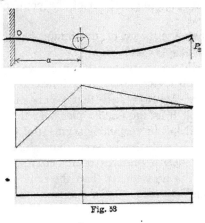

Fig. 53

The reaction at the supported end, being at present unknown, will be denoted by $P_2$, and moments will be taken on the right of any section $x$. From lack of symmetry, separate expressions must be written for segments on either side of $W$.

BETWEEN $W$ AND FIXED END.

$M_x = P_2(l-x) - W(a-x)$

$\dfrac{dv}{dx} = \dfrac{1}{EI}[P_2(lx - \tfrac{1}{2}x^2) - W(ax - \tfrac{1}{2}x^2) + C]$

$v = \dfrac{1}{EI}\left[P_2\left(\dfrac{lx^2}{2} - \dfrac{x^3}{6}\right) - W\left(\dfrac{ax^2}{2} - \dfrac{x^3}{6}\right) + Cx + C'\right]$

$\dfrac{dv}{dx} = 0$ when $x=0$; $\therefore C=0$.

$v=0$ when $x=0$; $\therefore C'=0$.

BETWEEN $W$ AND SUPPORTED END.

$M_x = P_2(l-x)$

$\dfrac{dv}{dx} = \dfrac{1}{EI}[P_2(lx - \tfrac{1}{2}x^2) + C'']$

$v = \dfrac{1}{EI}\left[P_2\left(\dfrac{lx^2}{2} - \dfrac{x^3}{6}\right) + C''x + C'''\right]$

If $x=a$, $\dfrac{dv}{dx}$ on left $= \dfrac{dv}{dx}$ on right.

If $x=a$, $v$ on left $= v$ on right.

$\therefore C'' = -W(a^2 - \tfrac{1}{2}a^2) = -\tfrac{1}{2}Wa^2;$

$C''' = W(-\tfrac{1}{2}a^3 + \tfrac{1}{6}a^3 + \tfrac{1}{2}a^3) = \tfrac{1}{6}Wa^3.$

If $x=l$, $v$ at $P_2 = 0$; $\therefore P_2(\tfrac{1}{2}l^3 - \tfrac{1}{6}l^3) - \tfrac{1}{2}Wa^2l + \tfrac{1}{6}Wa^3 = 0,$

or $\qquad P_2 = Wa^2(\tfrac{1}{2}l - \tfrac{1}{6}a) \div \tfrac{1}{3}l^3 = \dfrac{Wa^2}{2l^3}(3l-a).$

## RESTRAINED AND CONTINUOUS BEAMS.

If this value of $P_2$ is substituted in the above equations the desired expressions are obtained. Thus

$$M_x = \frac{Wa^2}{2l^3}(3l-a)(l-x) - W(a-x). \qquad M_x = \frac{Wa^2}{2l^3}(3l-a)(l-x).$$

$M_{max.}$, by inspection, when $x=0$, or $x=a$.

$$M_{max.} = \frac{Wa^2}{2l^2}(3l-a) - Wa = -\frac{W}{2l^2}a(l-a)(2l-a) \text{ at fixed end,}$$

and
$$= \frac{Wa^2}{2l^3}(l-a)(3l-a) \text{ at the weight.}$$

The point of contraflexure occurs between $W$ and the fixed end, where $M=0$, or

$$\frac{a^2}{2l^3}(3l-a)(l-x) - (a-x) = 0. \quad \therefore\ x = al\frac{2l-a}{2l(l+a) - a^2}.$$

The maximum deflection will be found where $\frac{dv}{dx} = 0$, on the right or left, according to the value of $a$.

The above beam may be regarded in the light of two equal continuous spans with $W$ on each, distant $a$ each side of the middle point of support.

In solving the more intricate problems in the flexure of beams, as well as those just treated, each equation of condition can be used but once in the same problem, and as many unknown quantities can be determined as there are independent equations of condition. The reactions and moments at the points of support are usually unknown, and must be found by the aid of such flexure equations as have just been used.

*Example.*—A bridge stringer which is continuous over two successive openings of 12 ft. each and carries a weight from the wheels of a wagon of 3,000 lb. at each side of and 3 ft. from the middle support will be horizontal over that support. Then $-M_{max.}$
$= -\frac{3,000}{2\cdot 12^2}\cdot 3\cdot 9\cdot 21 = -5,906.25$ ft.-lb. $+ M_{max.} = \frac{3,000}{2\cdot 12^3}\cdot 3^2\cdot 33\cdot 9$
$= 2,320.3$ ft.-lb. $P_2 = \frac{3,000}{2\cdot 12^3}\cdot 3^2\cdot 33 = 258$ lb. Reaction at middle support from both spans $= 2(3,000 - 258) = 5,484$ lb.

**109. Clapeyron's Formula, or the Three-moment Theorem for Continuous Loading.**—To find the reactions, shears, and bending moments for a horizontal uniform continuous beam loaded with $w_1$, $w_2$, $w_3$, etc., loads per running unit over the successive spans $l_1$, $l_2$, $l_3$, etc. Fig. 54. $P_0$, $P_1$, $P_2$, etc., denote

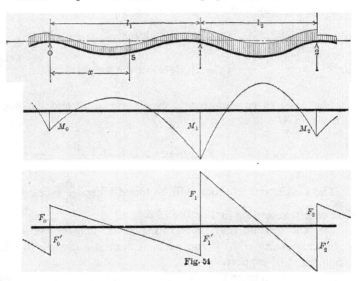

Fig. 54.

the unknown reactions; $M_0$, $M_1$, $M_2$, etc., the unknown bending moments at points of support, 0, 1, 2, etc.; $F_0$, $F_1$, $F_2$, etc., denote the shears immediately to the right of 0, 1, 2, etc.; while $F_1'$, $F_2'$, etc., denote the shears immediately to the left of the points of support 1, 2, etc.

The origin of coordinates is first taken at O and the supports are on a level. $+M$ makes the beam concave on the upper side. As positive shear acts upward at the left of any cross-section, $w$ is negative.

Consider the condition of equilibrium of the first span 0–1, or $l_1$, loaded throughout with $w_1$ per unit of length.

Take moments on the left side of and about a section S, distant $x$ from the origin O. The bending moment at S is

$$M = EI\frac{d^2v}{dx^2} = M_0 + F_0 x - \tfrac{1}{2}w_1 x^2. \quad \ldots \quad (1)$$

Let $i_0, i_1, i_2 \ldots i_n$ = tangent of inclination of the neutral axis at 0, 1, 2, ... N. Integrate (1) and transpose the constant of integration, $i_0$, to the left-hand member and thus obtain an expression for the difference in slope or inclination of the two tangents to the bent beam at O and S.

$$EI\left(\frac{dv}{dx} - i_0\right) = M_0 x + \tfrac{1}{2}F_0 x^2 - \tfrac{1}{6}w_1 x^3. \quad \ldots \quad (2)$$

When $x = l_1$, $\frac{dv}{dx} = i_1$, and hence

$$EI(i_1 - i_0) = M_0 l_1 + \tfrac{1}{2}F_0 l_1^2 - \tfrac{1}{6}w_1 l_1^3. \quad \ldots \quad (3)$$

Integrate (2) and determine constant as zero, because $v = 0$ when $x = 0$.

$$EI(v - i_0 x) = \tfrac{1}{2}M_0 x^2 + \tfrac{1}{6}F_0 x^3 - \tfrac{1}{24}w_1 x^4. \quad \ldots \quad (4)$$

Make $x = l_1$, then $v = v_1 = 0$,

and $\qquad -EI i_0 l_1 = \tfrac{1}{2}M_0 l_1^2 + \tfrac{1}{6}F_0 l_1^3 - \tfrac{1}{24}w_1 l_1^4,$

or $\qquad -EI i_0 = \tfrac{1}{2}M_0 l_1 + \tfrac{1}{6}F_0 l_1^2 - \tfrac{1}{24}w_1 l_1^3. \quad \ldots \quad (5)$

Eliminate $i_0$ by subtracting (5) from (3).

$$EI i_1 = \tfrac{1}{2}M_0 l_1 + \tfrac{1}{3}F_0 l_1^2 - \tfrac{1}{8}w_1 l_1^3. \quad \ldots \quad (6)$$

If the origin is taken at 1 instead of 0 an equation like (5) is obtained for the second span $l_2$, or

$$-EI i_1 = \tfrac{1}{2}M_1 l_2 + \tfrac{1}{6}F_1 l_2^2 - \tfrac{1}{24}w_2 l_2^3. \quad \ldots \quad (7)$$

Add (6) and (7), obtaining

$$0 = \tfrac{1}{2}M_0 l_1 + \tfrac{1}{2}M_1 l_2 + \tfrac{1}{3}F_0 l_1^2 + \tfrac{1}{6}F_1 l_2^2 - \tfrac{1}{8}w_1 l_1^3 - \tfrac{1}{24}w_2 l_2^3. \quad (8)$$

The unknown slopes have thus been eliminated. The next step is to remove either $M$ or $F$. Equation (1) must equal $M_1$, for $x=l_1$. Therefore

$$M_1 = M_0 + F_0 l_1 - \tfrac{1}{2}w_1 l_1^2, \quad \text{or} \quad F_0 = \frac{M_1 - M_0}{l_1} + \tfrac{1}{2}w_1 l_1.$$

In the same way, for second span, $F_1 = \dfrac{M_2 - M_1}{l_2} + \tfrac{1}{2}w_2 l_2$.

Substitute the values in (8) and obtain

$$0 = \frac{M_0 l_1}{2} + \frac{M_1 l_2}{2} + \frac{M_1 - M_0}{3}l_1 + \frac{w_1 l_1^3}{6} + \frac{M_2 - M_1}{6}l_2$$
$$+ \frac{w_2 l_2^3}{12} - \frac{w_1 l_1^3}{8} - \frac{w_2 l_2^3}{24}.$$

$$\therefore M_0 l_1 + 2M_1(l_1 + l_2) + M_2 l_2 = -\tfrac{1}{4}(w_1 l_1^3 + w_2 l_2^3), \quad . \quad . \quad (9)$$

which is Clapeyron's formula for pier moments for a continuous beam, with continuous load, uniform per span. Notice the symmetry of the expression. The negative sign to the second member indicates that the bending moments at points of support are usually negative.

*Example.*—Three spans, 30 ft., 60 ft., and 30 ft. in succession. Load on first and last 500 lb. per ft., on middle span 300 lb. per ft. No moment at either outer end. Then $M_0 = 0$. $M_1 = M_2$ by symmetry.

$2M_1 \cdot 90 + M_2 \cdot 60 = -\tfrac{1}{4}(500 \cdot 30^3 + 300 \cdot 60^3).$
$M_1 = -81,562\tfrac{1}{2}$ ft.-lb.    $F_0 = -2,718\tfrac{3}{4} + 7,500 = +4,781\tfrac{1}{4}$ lb.
$F_1' = +4,781\tfrac{1}{4} - 30 \cdot 500 = -10,218\tfrac{3}{4}$ lb.    $F_1 = 300 \cdot 30 = 9,000$ lb.
$\therefore P_0 = P_4 = 4,781\tfrac{1}{4}$ lb.;    $P_1 = P_2 = 10,218\tfrac{3}{4} + 9,000 = 19,218\tfrac{3}{4}$ lb.

The bending moment and shear at any point can now be readily determined.

If the two adjacent spans are equal and have the same load,

$$M_0 + 4M_1 + M_2 = -\tfrac{1}{2}wl^2. \quad . \quad . \quad . \quad . \quad (10)$$

If there are $n$ spans, $n-1$ equations can be written between $n+1$ quantities $M_0, M_1 \ldots M_n$. But if the beam is simply placed on the points of support, the extremities being unrestrained, $M_0 = 0$ and $M_n = 0$, and there remain $n-1$ equations

## RESTRAINED AND CONTINUOUS BEAMS.

to determine $M_1 \ldots M_{n-1}$. If the beam is fixed at the ends, the equations $i_0 = 0$ and $i_n = 0$ will complete the required number.

**110. Shears and Reactions.**—As the shear is the first derivative of the bending moment, § 56, from (1) is obtained

$$\frac{dM}{dx} = F = F_0 - w_1 x, \quad \ldots \ldots \quad (11)$$

as was to be expected, $+F$ acting upwards on the left of the section. A similar equation can be written for each span.

The reaction at any point of support will be equal to the shear on its right plus that on its left with the sign reversed. As the shear on its left is usually negative, the arithmetical sum of $F_n$ and $F_n'$ commonly gives the reaction.

A simple example may make the application plainer. Given two equal spans on three supports,

$w_1 = w_2 = w$. $M_0 = 0$, $M_2 = 0$. (10) gives $M_1 = -\frac{1}{8}wl^2$.
$F_0 = -\frac{1}{8}wl + \frac{1}{2}wl = \frac{3}{8}wl$; $\quad F_1' = \frac{3}{8}wl - wl = -\frac{5}{8}wl$.
$F_1 = \frac{1}{8}wl + \frac{1}{2}wl = \frac{5}{8}wl$; $\quad F_2' = \frac{5}{8}wl - wl = -\frac{3}{8}wl$.
$P_0 = \frac{3}{8}wl$; $\quad P_1 = (\frac{5}{8} + \frac{5}{8})wl = \frac{5}{4}wl$; $\quad P_2 = \frac{3}{8}wl$.

$$(5) \text{ gives } i_0 = -\frac{1}{EI}(0 + \tfrac{3}{48} - \tfrac{1}{24})wl^3 = -\frac{wl^3}{48EI}.$$

$$(6) \text{ gives } i_1 = \frac{1}{EI}(0 + \tfrac{3}{24} - \tfrac{1}{8})wl^3 = 0,$$

and the analogous equation for the second span is

$$i_2 = \frac{1}{EI}(-\tfrac{1}{16} + \tfrac{5}{24} - \tfrac{1}{8})wl^3 = \frac{wl^3}{48EI},$$

which differs from $i_0$ only in direction of slope.

$$(2) \text{ gives } EI\frac{dv}{dx} = -\frac{wl^3}{48} + \tfrac{3}{16}wlx^2 - \frac{wx^3}{6}.$$

$$(4) \text{ gives } EIv = -\frac{wl^3}{48}x + \tfrac{3}{48}wlx^3 - \frac{wx^4}{24}.$$

These equations determine the slope and deflection at each point. Putting $\dfrac{dv}{dx}=0$, there results $l^3-9lx^2+8x^3=0$, containing the root $x=l$, already known. Therefore divide by $l-x$ and obtain $l^2+lx-8x^2=0$, which is satisfied for $x=0.4215l$, the point of maximum deflection. The substitution of this value in the equation for $v$ will yield $v_{max}$.

From (1) $M=\tfrac{3}{8}wlx-\tfrac{1}{2}wx^2$.

If $M=0$, $\tfrac{3}{8}l-\tfrac{1}{2}x=0$, or $x=\tfrac{3}{4}l$, the point of contraflexure.
Differentiate $M$ and get $F=\tfrac{3}{8}wl-wx$.
If $F=0$, $x=\tfrac{3}{8}l$, the point of $+M_{max}$.
$\therefore M_{max.}=\tfrac{3}{8}wl\cdot\tfrac{3}{8}l-\tfrac{1}{2}w\cdot\tfrac{9}{64}l^2=\tfrac{9}{128}wl^2$.

*Example.*—If a uniformly loaded continuous beam covers five equal spans,

$$M_0+4M_1+M_2=-\tfrac{1}{2}wl^2=M_1+4M_2+M_3=M_2+4M_3+M_4$$
$$=M_3+4M_4+M_5.$$
$$M_0=0;\quad M_5=0.$$

Then $M_1=-\tfrac{4}{38}wl^2=M_4;\ M_2=-\tfrac{3}{38}wl^2=M_3$.

$F_0=\tfrac{15}{38}wl;\ F_1'=-\tfrac{23}{38}wl;\ F_1=\tfrac{19}{38}wl;\ F_2'=-\tfrac{19}{38}wl;\ F_2=\tfrac{19}{38}wl$, etc.

$P_0=\tfrac{15}{38}wl=P_5:\ P_1=\tfrac{43}{38}wl=P_4;\ P_2=\tfrac{34}{38}wl=P_3$.

The sum of the reactions must equal $5wl$.

**111. Coefficients for Moments and Shears.**—It has been found that the numerical coefficients for moments and shears at the points of support, when all spans are equal and the load is uniform throughout, may be tabulated easily for reference and use. Thus the values of $M$ and $F$ just obtained for the five equal spans can be selected from the lines marked V. The reactions are given by the *arithmetical* addition of the shears. The sum of the reactions must equal the total load. The shears at the two ends of any span differ by the whole load on the span, the shear at the right end being negative. The dashes represent the spans.

# RESTRAINED AND CONTINUOUS BEAMS.

### SHEAR AND REACTION COEFFICIENTS.

$$\text{I. } \tfrac{1}{2}\text{---}\tfrac{1}{2}wl;$$
$$\text{II. } \tfrac{3}{8}\text{---}\tfrac{5}{8} \quad \tfrac{5}{8}\text{---}\tfrac{3}{8}wl;$$
$$\text{III. } \tfrac{4}{10}\text{---}\tfrac{6}{10} \quad \tfrac{5}{10}\text{---}\tfrac{5}{10} \quad \tfrac{6}{10}\text{---}\tfrac{4}{10};$$
$$\text{IV. } \tfrac{11}{28}\text{---}\tfrac{17}{28} \quad \tfrac{15}{28}\text{---}\tfrac{13}{28} \quad \tfrac{13}{28}\text{---}\tfrac{15}{28} \quad \tfrac{17}{28}\text{---}\tfrac{11}{28};$$
$$\text{V. } \tfrac{15}{38}\text{---}\tfrac{23}{38} \quad \tfrac{20}{38}\text{---}\tfrac{18}{38} \quad \tfrac{19}{38}\text{---}\tfrac{19}{38} \quad \tfrac{18}{38}\text{---}\tfrac{20}{38} \quad \tfrac{23}{38}\text{---}\tfrac{15}{38};$$

etc., etc.

### PIER MOMENT COEFFICIENTS.

$$\text{II. } -\tfrac{1}{8}-wl^2;$$
$$\text{III. } -\tfrac{1}{10} \quad -\tfrac{1}{10}-wl^2;$$
$$\text{IV. } -\tfrac{3}{28} \quad -\tfrac{2}{28} \quad -\tfrac{3}{28}-;$$
$$\text{V. } -\tfrac{4}{38} \quad -\tfrac{3}{38} \quad -\tfrac{3}{38} \quad -\tfrac{4}{38}-;$$

etc., etc.

The rule for writing either table is as follows: For an even number of spans, the numbers in any horizontal line are obtained by multiplying the fraction above, in any *diagonal* row, both numerator and denominator, by *two*, and adding the numerator and denominator of the preceding fraction. Thus, in the first table, $\frac{2\times 6+5}{2\times 10+8}=\frac{17}{28}$, and in the second table, $\frac{2\times 1+1}{2\times 10+8}=\frac{3}{28}$, or $\frac{2\times 3+2}{2\times 38+28}=\frac{8}{104}$. For an odd number of spans, add the two preceding fractions in the same diagonal row, numerator to numerator and denominator to denominator. Thus, $\frac{13+5}{28+10}=\frac{18}{38}$. The denominators agree in both tables. A recollection of two or three quantities will enable one to write all the others.

*Example.*—Continuous beam of five equal spans, each $l$, carrying $w$ per foot. Where and what is the max. $+M$ in second span? Shear changes sign at $\frac{20}{38}l$ from left end of span. If this span were independent, $+M$ at that point would be $\tfrac{1}{2}wl^2 \cdot \frac{20 \cdot 18}{19 \cdot 19}=\frac{360}{361} \cdot \frac{wl^2}{8}$. The negative or subtractive moment is $(\tfrac{3}{38}+\tfrac{1}{3}\cdot\tfrac{18}{38})wl^2$. The difference between these values is $+M_{max}$.

A more general investigation will produce equations which are of great practical value in the solution of problems concerning continuous bridges, swing-bridges, etc., as follows:

**112. Three-moment Theorem for a Single Weight.**—O is the origin, Fig. 55; the supports are at distances $h$ below the axis of X. A single weight $W_n$ is distant $kl_n$ from O on the span $l_n$, $k$ being a fraction, less than unity, of the span in which $W$ is situated.

The moment at section S beyond $W_n$ will be, as in the former discussion,

$$M_x = M_n + F_n x - W_n(x - kl_n). \quad \cdots \quad (1)$$

If $x = l_n$, $M_x = M_{n+1}$, and from (1)

$$F_n = \frac{M_{n+1} - M_n}{l_n} + W_n(1 - k). \quad \cdots \quad (1a)$$

For an unloaded span, $W = 0$, and $F_m = \dfrac{M_{m+1} - M_m}{l_m}$.

For the shear on the left of a section at the right end of the $n$th span,

$$F_{n+1}' = F_n - W_n = \frac{M_{n+1} - M_n}{l_n} - W_n k.$$

For an unloaded span, $W = 0$, and

$$F_m' = \frac{M_m - M_{m-1}}{l_{m-1}}.$$

As $F_m'$ is the shear at left of support $m$, and $F_m$ is the shear at right of the same support; the reaction there will be the sum of $F_m$ and $-F_m'$ or

$$P_m = \frac{M_{m+1} - M_m}{l_m} + \frac{M_{m-1} - M_m}{l_{m-1}}.$$

To get values of $i$ and $v$ for span $l_n$ it is necessary to write separate equations for the two segments into which $W_n$ divides the span. Equation (1) applies to the right segment, and by omitting the last term the equation for the left segment is obtained. Equate $\frac{M}{EI}$ to $\frac{d^2v}{dx^2}$ and integrate between the limits of o and

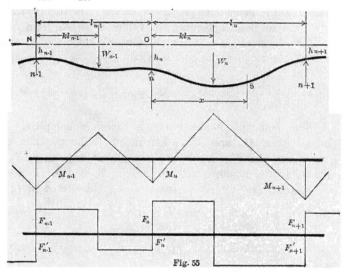

Fig. 55

$kl_n$ on the left and between $kl_n$ and $x$ on the right. Then on the left

$$EI(i_{kl_n} - i_n) = M_n \int_0^{kl_n} dx + F_n \int_0^{kl_n} x\,dx, \quad \ldots \quad (2)$$

and on the right

$$EI\left(\frac{dv}{dx} - i_{kl_n}\right) = M_n \int_{kl_n}^{x} dx + F_n \int_{kl_n}^{x} x\,dx - W_n \int_{kl_n}^{x} (x - kl_n)\,dx. \quad (3)$$

The sum of (2) and (3) gives the change of slope between the support $n$ and the section S.

$$EI\left(\frac{dv}{dx} - i_n\right) = M_n \int_0^{x} dx + F_n \int_0^{x} x\,dx - W_n \int_{kl_n}^{x} (x - kl_n)\,dx$$
$$= M_n x + \tfrac{1}{2} F_n x^2 - \tfrac{1}{2} W_n (x - kl_n)^2. \quad \ldots \quad (4)$$

Integrate the last equation between the same limits as before.

$$EI(v - i_n x - h_n) = \tfrac{1}{2} M_n x^2 + \tfrac{1}{6} F_n x^3 - \tfrac{1}{6} W_n (x - k l_n)^3, \quad (5)$$

which is the general equation of the curve of the neutral axis, the term in $W$ disappearing for values of $x$ less than $k l_n$.

If $x = l_n$, $v = h_{n+1}$. If the value of $F_n$ from (1a) is inserted in (5), the slope at support $n$ is

$$i_n = \frac{h_{n+1} - h_n}{l_n} - \frac{1}{6EI} [2 M_n l_n + M_{n+1} l_n + W_n l_n^2 (2k - 3k^2 + k^3)]. \quad (6)$$

The equation of the curve is therefore completely determi ed when $M_n$ and $M_{n+1}$ are known. The equation of this curve, between $W_n$ and the $n+1$th support is given by (5), and the tangent of its angle with the axis of X by (4). If the value of $F_n$ from (1a) and of $i_n$ from (6) are substituted in (4), and $x = l_n$, $\dfrac{dv}{dx}$ will be the tangent $i_{n+1}$ at $n+1$,

or $\quad i_{n+1} = \dfrac{h_{n+1} - h_n}{l_n} + \dfrac{1}{6EI} [M_n l_n + 2 M_{n+1} l_n + W_n l_n^2 (k - k^3)].$

Remove the origin from O to N, and derive an expression for $i_n$ by diminishing the indices.

$$i_n = \frac{h_n - h_{n-1}}{l_{n-1}} + \frac{1}{6EI} [M_{n-1} l_{n-1} + 2 M_n l_{n-1} + W_{n-1} l_{n-1}^2 (k - k^3)].$$

Equate with (6) and transpose.

$$M_{n-1} l_{n-1} + 2 M_n (l_{n-1} + l_n) + M_{n+1} l_n$$
$$= 6EI \left( \frac{h_{n-1} - h_n}{l_{n-1}} + \frac{h_{n+1} - h_n}{l_n} \right) - W_{n-1} l_{n-1}^2 (k - k^3)$$
$$- W_n l_n^2 (2k - 3k^2 + k^3), \quad (7)$$

which is the most general form of the three-moment theorem for a girder of constant cross-section. In using the theorem the following factors may be of service:

$$k - k^3 = k(1 - k)(1 + k); \quad 2k - 3k^2 + k^3 = k(1 - k)(2 - k).$$

## RESTRAINED AND CONTINUOUS BEAMS.

Pier moments are usually negative and the end moments zero. When the supports are on a level, $h_1 = h_2$, etc., and the term containing $EI$ disappears.

Any reaction $P_n = F_n - F_n'$;

$$\therefore P_n = \frac{M_{n+1} - M_n}{l_n} + \frac{M_{n-1} - M_n}{l_{n-1}} + W_n(1-k) + W_{n-1}k.$$

It should be borne in mind that all the preceding deflection formulas are derived on the assumption that the deflection is due entirely to bending moment, the deformation from shear being neglected. In a solid beam the amount of deflection due to shear is very small, but such is not the case with a truss. In a truss the deflection due to the deformation of the web may in some cases amount to half the total deflection. For this reason the deflection formulas and the three-moment theorem should not be applied to trusses.

*Example.*—Three-span continuous girder carrying loads as shown. Supports on a level.

$2 \cdot 150 M_1 + 100 M_2 = -2{,}000 \cdot 2{,}500 \cdot \frac{3}{5} \cdot \frac{2}{5} \cdot \frac{8}{5} - 1{,}000 \cdot 10{,}000 \cdot \frac{2}{10} \cdot \frac{8}{10} \cdot \frac{18}{10}$
$\qquad\qquad\qquad - 3{,}000 \cdot 10{,}000 \cdot \frac{6}{10} \cdot \frac{4}{10} \cdot \frac{14}{10}.$
$100 M_1 + 2 \cdot 175 M_2 = -1{,}000 \cdot 10{,}000 \cdot \frac{2}{10} \cdot \frac{8}{10} \cdot \frac{12}{10} - 3{,}000 \cdot 10{,}000 \cdot \frac{6}{10} \cdot \frac{4}{10} \cdot \frac{16}{10}.$
$300 M_1 + 100 M_2 = -14{,}880{,}000, \qquad 100 M_1 + 350 M_2 = -13{,}440{,}000.$
$M_1 = -40{,}670$ ft.-lb. $\qquad M_2 = -26{,}780$ ft.-lb.

$$P_0 = \frac{-40{,}670}{50} + \frac{2{,}000 \cdot 20}{50} = -13 \text{ lb.}$$

$$P_1 = \frac{-26{,}780 + 40{,}670}{100} + \frac{40{,}670}{50} + \frac{1{,}000 \cdot 8}{10} + \frac{3{,}000 \cdot 4}{10} + \frac{2{,}000 \cdot 3}{5} = +4{,}152 \text{ lb.}$$

$$P_2 = \frac{+26{,}780}{75} + \frac{-40{,}670 + 26{,}780}{100} + \frac{1{,}000 \cdot 2}{10} + \frac{3{,}000 \cdot 6}{10} = +2{,}218 \text{ lb.}$$

$$P_3 = \frac{-26{,}780}{75} = -357 \text{ lb.}$$

$-13 + 4{,}152 + 2{,}218 - 357 = 2{,}000 + 1{,}000 + 3{,}000 = 6{,}000.$

*Examples.*—1. A brick wall 16 in. thick, 12 ft. high, and 32 ft. long, weighing 108 lbs. per cubic ft., is carried on a beam supported by four columns, one at each end, and one 8 ft. from each end. Find $M$ at the two middle columns and the reactions.

$$M = -31{,}104 \text{ ft.-lb.}; \quad P_1 = 3{,}024 \text{ lb.}; \quad P_2 = 24{,}624 \text{ lb.}$$

2. Two successive openings of 8 ft. each are to be spanned. Which will be stronger for a uniform load, two 8 ft. joists end to end or one 16 ft. long? Find their relative stiffness.

3. A beam of three equal spans carries a single weight. What will be the reactions and their signs at the third and fourth points of support when $W$ is in the middle of the first span?

$$-\tfrac{3}{26}W; \; +\tfrac{1}{40}W.$$

4. A beam loaded with 50 lb. per foot rests on two supports 15 ft. apart and projects 5 ft. beyond at one end. What additional weight must be applied to that end to make the beam horizontal at the nearer point of support?  $156\tfrac{1}{4}$ lb. at the end, or $312\tfrac{1}{2}$ lb. distributed.

5. A beam of two equal spans on level supports carries a single load, $W$, in the left span. Prove that

$$P_1 = \tfrac{1}{4}W(4 - 5k + k^3);$$
$$P_2 = \tfrac{1}{2}W(3k - k^3);$$
$$P_3 = -\tfrac{1}{4}W(k - k^3).$$

# CHAPTER VIII.

## PIECES UNDER TENSION.

**113. Central Pull.**—If the resultant tension $P$ acts along the axis of the piece, the stress may be considered as uniformly distributed on the cross-section $S$. If, then, $f$ is the maximum safe working stress per square inch for the kind of load which causes $P$, Fig. 56,

$$P = fS, \quad \text{or} \quad S = \frac{P}{f}$$

for the necessary section which need not be exceeded throughout that portion of the piece where the above conditions apply. Changes due to connections will require a larger section.

If owing to lack of uniformity in the material or the direct application of $P$ at the end of a wide bar to a limited portion only of the width the stress may not be considered as uniformly distributed at a particular cross-section, injurious stress may be prevented by taking the mean stress $f$ at a smaller value and obtaining a larger cross-section.

If there is lack of homogeneity, or two materials are used together, or two or more bars work side by side, those fibres which offer the greatest resistance to stretching will be subject to the greatest stress. Fortunately the slight yielding and bending of connecting parts tend to restore equality of action.

A long tension member has a much greater resisting power against shock than a short one of equal strength per square inch. See § 20.

**114. Eccentric Pull.**—If the variation of stress on a cross-section is due to the fact that the line of action of the applied force does not traverse the centre of figure of the cross-section $S$, the force $P'$ that can be imposed without causing a unit stress greater than $f$ at any point in the section is less than $P$ of the preceding formula, and depends upon the perpendicular distance $y_0$ of the action line of $P$ from the centre of $S$.

For safe stresses, which must lie well within the elastic limit, the unit stress is proportional to the stretch, and plane cross-sections of the bar before the force is applied are assumed to remain plane after the bar is stretched. It is impossible to detect experimentally that this assumption is not true. Were the plane sections to become even slightly warped, the cumulative warpings of successive sections in a long bar ought to become apparent to the eye. No reference is intended here to local distortion preceding failure.

If the stress on any section is not uniform and the successive sections remain plane, they must be a little inclined to one another.

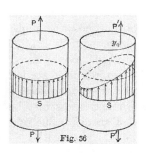
Fig. 56

The stress on any cross-section $S$ must therefore vary uniformly in the direction of the deviation of the action line of $P'$ from the centre, (Fig. 56), and be constant on lines at right angles to that deviation. The force, $P'$, acting at a distance, $y_0$, from the axis of the tie is equivalent to the same force, $P'$, at the centre and a moment, $P'y_0$. The stress on each particle of any section may therefore be divided into two parts, that due to direct tension and that due to the moment or to bending. The former is uniform across the section and is

$$f_t = \frac{P'}{S}.$$

The latter is zero at the centre of gravity of the section and a maximum at the edge toward which the load deviates, where it is

$$f_b = \frac{My_1}{I} = \frac{P'y_0y_1}{Sr^2}.$$

$$f = f_t + f_b = \frac{P}{S}\left(1 + \frac{y_0y_1}{r^2}\right) = f_t\left(1 + \frac{y_0y_1}{r^2}\right).$$

Note that $f_t$ is the mean stress, which is always the existing stress at the centre of gravity of the cross-section.

*Example.*—A square bar 1 in. in section carries 6,000 lb. tension. The centre of the eye at the end is $\frac{1}{4}$ in. out of line. Then $f = 6{,}000(1 + \frac{1}{4} \cdot \frac{1}{2} \cdot 12) = 15{,}000$ lb. per sq. in., $2\frac{1}{2}$ times the mean and probably the intended stress.

A bar which is not perfectly straight before tension is applied to it tends to straighten itself under a pull, but the stress will not become uniform on a cross-section. The bar is weaker in the ratio of $f_t$ to $f$, as it might carry $fS$ if the force were central, but now can safely carry only $f_tS$. If a thrust is applied to a bent bar, there is a tendency to increased deviation from a straight line, and to an increase in the variation of stress.

It is seen from the example above that a small deviation $y_0$ will have a decided effect in increasing $f$ for a given $P'$, or in diminishing the allowable load for a given unit stress. Herein may be the explanation of some considerable variations of the strength of apparently similar pieces under test; and, on account of such effect, added to other reasons, allowable working stresses may well be and are reduced below what otherwise might be used.

**115. Curved Piece in Bending.**—Fig. 57A shows a small length of a piece whose axis is a plane curve and whose section is symmetrical about the plane of curvature. An eccentrically applied force $P$, which is called $+$ if compressive and $-$ if tensile, acts upon the section, being replaced by the same force at the centre of gravity and a moment $M$. Because of the bending moment the radial plane AB will assume some such position as A'B', if sections plane before flexure are assumed to remain plane after flexure as with straight beams, and the deformation of any fibre will be proportional to its distance from the neutral axis; but as the original length of all fibres was not the same, the *unit* deformation of a fibre will *not* be proportional to its distance from the neutral axis and hence the unit stresses on the section will not vary uniformly.

Let the length of an unstressed fibre distant $y$ from the gravity axis be $l$ and let its change of length under stress be $\delta l$. Then $\delta l = \delta l_0 + y\delta\theta$ and

$$\lambda = \frac{\delta l}{l} = \frac{\delta l_0}{l_0} \cdot \frac{l_0}{l} + \frac{y\delta\theta}{\theta} \cdot \frac{\theta}{l}.$$

From the figure $\dfrac{l_0}{l} = \dfrac{\rho}{\rho+y}$, $\theta = \dfrac{l}{\rho+y}$, and $\dfrac{\delta l_0}{l_0} = \lambda_0$; hence

$$\lambda = \lambda_0 \frac{\rho}{\rho+y} + \frac{y}{\rho+y} \cdot \frac{\delta\theta}{\theta} = \lambda_0 + \left(\frac{\delta\theta}{\theta} - \lambda_0\right)\frac{y}{\rho+y}. \quad \ldots \quad (1)$$

If $p$ is the unit stress at a distance $y$ from the gravity axis, $p = E\lambda$, and that the internal stresses on the section may balance the external forces

$$P = \int p\, dS \quad \text{and} \quad M = \int p\, dS \cdot y, \quad \ldots \quad (2)$$

where $z\,dy$ is represented by $dS$. Multiply (1) by $E$ and substitute in (2), representing $\delta\theta/\theta$ by $\omega$.

$$P = E\lambda_0 \int dS + E(\omega - \lambda_0) \int \frac{y\,dS}{\rho+y},$$

$$M = E\lambda_0 \int y\,dS + E(\omega - \lambda_0) \int \frac{y^2\,dS}{\rho+y}.$$

Let $\displaystyle\int \frac{y\,dS}{\rho+y} = -kS.$

Then $\displaystyle\int \frac{y^2\,dS}{\rho+y} = \int \left(y - \frac{\rho y}{\rho+y}\right) dS = -\rho \int \frac{y\,dS}{\rho+y} = k\rho S,$

since $\int y\,dS = 0.$

$$P = ES\{\lambda_0 - (\omega - \lambda_0)k\},$$
$$M = ES(\omega - \lambda_0)k\rho,$$
$$(\omega - \lambda_0) = \frac{M}{ESk\rho}, \qquad \lambda_0 = \frac{P}{ES} + \frac{M}{ES\rho}.$$

Substitute these values in (1) and multiply by $E$:

$$p = \frac{P}{S} + \frac{M}{\rho S} + \frac{M}{k\rho S} \cdot \frac{y}{\rho+y}, \quad \ldots \quad (3)$$

in which

$$k = -\frac{1}{S} \int \frac{y\,dS}{\rho+y}.$$

When $\rho$ is large compared with the depth of the cross-section

## PIECES UNDER TENSION.

$$k\rho S = \int \frac{y^2 dS}{\rho+y} = \frac{1}{\rho}\int \frac{y^2 dS}{1+y/\rho} = \frac{1}{\rho}\int y^2 dS \text{ nearly} = \frac{I}{\rho}, \text{ or } k = \frac{I}{\rho^2 S}$$

nearly; and if $\rho = \infty$, (3) becomes $p = \frac{P}{S} + \frac{My}{I}$, the stress in a straight beam.

The value of $k$ can be determined by integration for simple forms of cross-section. If $h$ is the depth of the piece, for the Rectangle, $k = -1 + \frac{\rho}{h}\log\frac{\rho+\frac{1}{2}h}{\rho-\frac{1}{2}h} = \frac{1}{3}\left(\frac{h}{2\rho}\right)^2 + \frac{1}{5}\left(\frac{h}{2\rho}\right)^4 + \frac{1}{7}\left(\frac{h}{2\rho}\right)^6 + \ldots;$

Ellipse and Circle,

$$k = -1 + 2\frac{2\rho}{h}\left(\frac{2\rho}{h} - \sqrt{\left(\frac{2\rho}{h}\right)^2 - 1}\right) = \frac{1}{4}\left(\frac{h}{2\rho}\right)^4 + \frac{1}{8}\left(\frac{h}{2\rho}\right)^4 + \frac{5}{64}\left(\frac{h}{2\rho}\right)^6 + \ldots$$

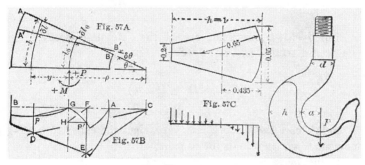

FIG. 57.

For irregular forms $k$ can be determined graphically as shown in Fig. 57B. The construction follows directly from the general expression for $k$. GP is drawn parallel to CD, and FH to CE. (Area GP'A − Area BPG) ÷ area BDEA = $k$. Hence for all similar sections having the centre of curvature at a similar distance $k$ is the same.

**116. Hooks.**—The load on a hook, as commonly made, is applied through the centre of curvature, whence $M = P\rho$. If this value is substituted in (3) and $P$ is made negative because the force is tensile, and if $p$ and $y$ are replaced by $f$ and $-y_1$, the capacity of the hook is found to be

$$P = \frac{fakS}{y_1},$$

in which $a = \rho - y_1 =$ the radius of the hook opening. The distribution of stress on the section is as shown.

*Example.*—Fig. 57C shows the usual proportions of crane-hook sections. The area is $0.426h^2$, $y_1$ is $0.435h$, and the round from which it can be forged is $d=0.74h$.

For $a \div h=$    0.4      0.6      0.8      1.0
      $k=$    0.120    0.075    0.055    0.040

**117. Combined Tie and Beam.**—If to a tension member transverse forces are applied, or if it is horizontal and its weight is of importance, the unit tensile stress on the convex edge, due to the maximum bending moment, must be added to the unit stress at that point due to the direct pull. As in § 114,

$$f_b = \frac{(M_{max.})y_1}{I}, \quad \text{and} \quad f_t = \frac{P}{S}.$$

But $f = f_b + f_t$ must not exceed the safe unit tension, and the needed section is, since $I = Sr^2$,

$$S = \frac{1}{f}\left(\frac{(M_{max.})y_1}{r^2} + P\right).$$

In this case the sections may vary, since the external bending moment $M$ varies from point to point.

If the piece is rectangular in section, as with timber, the formula may be written

$$f = \frac{6M}{bh^2} + \frac{P}{bh}, \quad \text{or} \quad b = \frac{1}{fh}\left(\frac{6M}{h} + P\right).$$

In practical calculation of such a rectangular section, if $h$ is assumed, it is sufficient to compute the breadth to carry $M$ and add enough breadth to carry $P$, when the combined section will have exactly $f$ at the edge.

*Example.*—A rectangular wooden beam of 12 ft. span carries a single weight of 3,000 lb. at the quarter span, and, as part of a truss, resists a pull of 20,000 lb. If $f = 1,000$ lb., what should be the section under the weight? $M_{max.} = \frac{3,000 \cdot 3 \cdot 9 \cdot 12}{12} = 81,000$ in.-lb. $\frac{81,000 \cdot 6}{1,000} = bh^2 = 486$. If $h = 12$, $b = 3.37$. Also, $\frac{20,000}{1,000 \cdot 12} = 1.67$. Entire breadth $= 3.37 + 1.67 = 5.04$. Section $= 5 \times 12$ in. The same result is obtained by the formula

$$b = \frac{1}{12 \cdot 1,000}\left(\frac{6 \cdot 81,000}{12} + 20,000\right).$$

If the tensile stress in a combined tie and beam is small as compared with the stress due to flexure the above solution is accurate enough, but when the tensile stress is large the error is considerable. The transverse forces acting on the beam cause it to deflect and consequently the line in which the direct pull acts does not pass through the centre of gravity of each section of the beam and a bending moment is produced which partially counteracts the bending moment due to the transverse forces. The bending moment at any section then is

$$M - Pv = \frac{f_b I}{y_1},$$

in which $v$ is the deflection of the tie-beam at the section where $f_b$ is found. By referring to Chapter VI it is seen that the deflection of a beam may be written

$$v = k\frac{f_b l^2}{E y_1},$$

where $k$ is a constant depending upon the way in which the beam is loaded. If it is assumed that $v$ is proportional to $f_b$ in the case under consideration as well as in the case of a beam not subjected to tension, there results

$$f_b = \frac{M y_1}{I + k\dfrac{P l^2}{E}}.$$

For a beam supported at both ends and uniformly loaded $k = \frac{5}{48}$; carrying a single load at the centre, $k = \frac{1}{12}$; carrying a single load at the quarter-point, $k = \frac{1}{16}$. If the ends are free to turn and the transverse load is uniformly distributed or is a single load near the centre, $k$ may be taken as $\frac{1}{10}$.

An expression for the fibre stress in a horizontal steel eye-bar bending under its own weight is readily derived from the last equation. Since a bar of steel one square inch in area and one foot long weighs 3.4 lb. and $E$ is 29,000,000, there results when all dimensions are in inches:

$$f_b = \frac{4,900,000\,h}{f_t + 23,000,000\left(\dfrac{h}{l}\right)^2}.$$

*Example.*—An eye-bar $8\times 1$ in., 25 ft. long, carries 100,000 lb. tension. $f_t = 12,500$ lb. per sq. in. If the effect of the deflection is neglected, $f_b = 2,390$ lb. Taking the deflection into account by the last formula, $f_b = 1,360$ lb.

**118. Action Line of P Moved towards the Concave Side.**—It will be economical, if it can be done, in a member having such compound action, to move the line on which $P$ acts towards the concave side. If there are bending moments of opposite signs at different points of the length, or at the same point at different times, such adjustment cannot be made. If $y_0$ is made equal to $\tfrac{1}{2}(M_{max.}) \div P$, one-half of the bending moment will be annulled at the point where $M_{max.}$ exists, and at the point of no bending moment from transverse forces an equal amount of bending moment will be introduced. The unit stresses on the extreme fibres at the two sections will be the same, but reversed one for the other.

*Example.*—A horizontal bar, $6\times 1$ in. section and 15 ft. long has a tension of 45,000 lb. It carries 100 lb. per foot uniformly distributed. $M_{max.} = \dfrac{100\cdot 15\cdot 15\cdot 12}{8} = 33,750$ in.-lb. $\therefore y_0$ may be made $\tfrac{3}{8}$ in. Then $f$ from $M_{max.} = \dfrac{33,750\cdot 6}{36} = \pm 5,625$ lb. on either edge. But $f$ from $P = \dfrac{45,000}{6}\left(1 \pm \dfrac{3\cdot 3\cdot 12}{8\cdot 6^2}\right) = 7,500(1\pm\tfrac{3}{8}) = 7,500 \pm 2,812.5$. Stress at top at ends and at bottom at middle $10,312\tfrac{1}{2}$ lb.; at bottom at ends and at top at middle $4,687\tfrac{1}{2}$ lb.

The extreme fibre stress from bending moment of the load varies as the ordinates to a parabola; that from $Py_0$ is constant. A rectangle superimposed on a parabolic segment will show the resultant fibre stress at each section.

**119. Connecting-rod.**—If a bar oscillates laterally rapidly, as does a connecting-rod on an engine, or a parallel rod on locomotive drivers, there are forces developed due to the acceleration, and at certain positions of the bar these forces are transverse and cause bending. When a *parallel rod* is in its highest or lowest position the centrifugal force, due to the circular motion of every part of the rod, is acting at right angles to the bar, which is then subjected to a uniformly distributed transverse load of

$$\frac{wv^2}{gr} = \frac{w}{g}4\pi^2 n^2 R,$$

in which $n$ is the number of revolutions per second, $R$ the radius of the crank, $w$ the weight of unit length of bar and $g$ the acceleration due to gravity.

The rotating end of a *connecting-rod* at two points in its path is under the influence of a transverse force whose intensity is obtained by the above formula, but as there can be no transverse force at the sliding end due to acceleration, the rod as a whole may be considered to be acted upon by transverse forces varying uniformly from $wv^2 \div gr$ at one end to zero at the other. The maximum fibre stress due to such loading may be found and added to the tensile or compressive stress caused by the pull or thrust along the rod. An I-shaped section is suitable for such members. Owing to the rapid variations and alternation of stress, the maximum unit stress should be small. Mass is disadvantageous in such rods.

**120. Tension and Torsion.**—A tension bar may be subjected to torsion when it is adjusted by a nut at the end, or by a turnbuckle. The moment of torsion will give rise to a unit shear at the extreme fibre, for a round rod, of $q_1 = T \div 0.196 d^3$ by § 82, or at the middle of the side, for a square rod, of $q_1 = T \div 0.208 h^3$ by § 84, either of which, combined with $f = P \div S$, the tensile stress, will give $p_1 = \frac{1}{2}f + \sqrt{(\frac{1}{4}f^2 + q^2)}$. See § 86.

*Example.*—A round bar, 2 in. diameter, to be adjusted to a pull of 10,000 lb. per square inch, calls for the application to the turnbuckle of 200 lb. with an arm of 30 in., one-half of which moment may be supposed to affect either half of the rod. If the turnbuckle is near one end, the shorter piece will experience the greater part of the moment. $q = \frac{3,000 \cdot 2 \cdot 7}{22 \cdot 1^3} = 1,910$ lb. The maximum unit tension on the outside fibres of the rod will be $5,000 + \sqrt{(5,000^2 + 1,910^2)} = 10,350$ lb.

**121. Tension Connections.**—If a tension member is spliced, or is connected at its ends to other members by rivets, the splice should be so made or the rivets should be so distributed across the section as to secure a uniform distribution of stress. An angle-iron used in tension should be connected by both flanges if the whole section is considered to be efficient. One or more rivet-holes must be deducted in calculating the effective section, depending on the spacing of the rivets. See § 185. If the stress is not uniformly distributed on the cross-section, the required size will be found by § 114, $S = \dfrac{P'}{f}\left(1 + \dfrac{y_0 y_1}{r^2}\right)$.

Transverse bolts and bolt-holes are similar to rivets and rivet-holes.

Timbers may be spliced by clamps with indents, and by scarfed joints, in which cases the net section is much reduced; so that timber, while resisting tension well, is not economical for ties, on account of the great waste by cutting away. However, where the tie serves also as a beam, timber may be very suitable.

**122. Screw-threads and Nuts.**—If a metal tie is secured by screw-threads and nuts, the section at the bottom of the thread should be some 15 per cent. larger than the given tension would require, to allow for the local weakening caused by cutting the threads. Bars are often upset or enlarged at the thread to give the necessary net section and thus save the material which would be needed for an increase of diameter throughout the length of the rod.

To avoid stripping the thread, the cylindrical surface, whose area is the circumference at the bottom of the thread multiplied by the effective thickness of the nut, should be, when multiplied by the safe unit shear, at least equal to the net cross-section of the rod multiplied by the safe unit tension. $2\pi R_1 \cdot t \cdot q_1 = \pi R_1^2 f$, or $f R_1 = 2 q_1 t$. As $q_1$ is usually taken less than $f$; as with a square thread only half of the thickness is effective, and with a standard V thread quite a portion of the thickness must be deducted, nuts are usually given a thickness nearly or quite equal to the net diameter of the rod. Heads of bolts may be materially thinner.

# CHAPTER IX.

## COMPRESSION PIECES—COLUMNS, POSTS, AND STRUTS.

**123. Blocks in Compression.**—If the height of the piece is quite small as compared with either of its transverse dimensions, and the load upon it is centrally imposed, the load or force $P$ may reasonably be considered as uniformly distributed over the cross-section $S$, and the unit stress $f$ upon each square inch of section will be given by the formula

$$P = fS, \quad \text{or} \quad f = \frac{P}{S},$$

as is the case with any tension member when the force is centrally applied.

**124. Load not Central.**—So also when the action line of the resultant load cannot be considered as central, but deviates from the axis of the piece a distance $y_0$, the force $P$ can be replaced by the same force acting in the axis and a couple or moment $Py_0$, which moment must be resisted at every cross-section by a uniformly varying stress, forming a resisting moment exactly like that found at a section of a beam. Compare Fig. 56, and change tension to compression.

If $f_c$ is the uniform unit stress due to a central load of $P$, and if $f_b$ is the unit stress on the extreme fibre lying in the direction in which $y_0$ is measured, the latter stress being due to the moment $Py_0$,

$$f_c = \frac{P}{S}; \quad f_b = \frac{Py_0 y_1}{I} = \frac{P}{S}\frac{y_0 y_1}{r^2}.$$

138        STRUCTURAL MECHANICS.

The greatest unit stress on the section is

$$f = f_c + f_b = \frac{P}{S}\left(1 + \frac{y_0 y_1}{r^2}\right) = f_c\left(1 + \frac{y_0 y_1}{r^2}\right).$$

The load that such a piece will carry is

$$P = \frac{fS}{1 + \frac{y_0 y_1}{r^2}}.$$

By comparison with the formula of the preceding section it will be seen that the piece, when the load is eccentric, is weaker in the ratio

$$f_c : f_c + f_b = 1 : 1 + \frac{y_0 y_1}{r^2}.$$

The values of $y_1 S \div I$ or $y_1 \div r^2$ are given below for some of the common sections of columns, $y_1$ being measured in the direction $h$.

|  | $I$ | $y_1$ | $S$ | $y_1 \div r^2$ |
|---|---|---|---|---|
| Rectangle........ | $\dfrac{bh^3}{12}$ | $\tfrac{1}{2}h$ | $bh$ | $\dfrac{6}{h}$ |
| Square. ......... | $\dfrac{h^4}{12}$ | $\tfrac{1}{2}h$ | $h^2$ | $\dfrac{6}{h}$ |
| Circle........... | $\dfrac{\pi d^4}{64}$ | $\tfrac{1}{2}d$ | $\tfrac{1}{4}\pi d^2$ | $\dfrac{8}{d}$ |
| Hollow rectangle... | $\dfrac{bh^3 - b'h'^3}{12}$ | $\tfrac{1}{2}h$ | $bh - b'h'$ | $\dfrac{6h(bh - b'h')}{bh^3 - b'h'^3}$ |
| Hollow circle...... | $\dfrac{\pi(d^4 - d'^4)}{64}$ | $\tfrac{1}{2}d$ | $\tfrac{1}{4}\pi(d^2 - d'^2)$ | $\dfrac{8d}{d^2 + d'^2}$ |

**125. The Middle Third.**—The mean or average unit stress is always found at the centre of gravity of the cross-section. On sections symmetrical about the neutral axis the fibre stress can be zero at one edge only when the fibre stress on the opposite edge is twice the mean, that is, when

$$f_b = f_c \quad \text{or} \quad y_0 = \frac{r^2}{y_1}.$$

## COMPRESSION PIECES. 139

This condition is satisfied for a rectangle when $y_0 = \frac{1}{6}h$. Hence for a rectangular section in masonry the centre of pressure must not deviate from the centre of figure more than one-sixth of the breadth in either direction, if the unit stress at the more remote edge is not to be allowed to become zero or tension. As masonry joints are supposed in many cases not to be subjected to tension in any part, the above statement is equivalent to saying that the centre of pressure or line of the resultant thrust must always lie within the *middle third* of any joint.

Likewise, for two cylindrical blocks in end contact, the centre of pressure should fall within the *middle fourth* of the diameter if the pressure is assumed to be uniformly varying and it is not permissible to have the joint tend either to open or to carry tension at the farther edge.

The unit pressure at the most pressed edge of a rectangle can be found for any deviation $y_0$.

1st. When the stress is over the whole joint, as before,

$$f = \frac{P}{S}\left(1 + \frac{y_0 y_1}{r^2}\right).$$

2d. When compression alone is possible, and only a part of the surface of the joint is under stress. The distance from the most pressed edge to the action line of $P$ is $\frac{1}{2}h - y_0$. The entire pressed area $= 3(\frac{1}{2}h - y_0)b$, since the ordinates representing stresses make a wedge whose length along $y$ is three times the distance of $P$ from the most pressed edge.

$$P = \tfrac{1}{2}f \cdot 3(\tfrac{1}{2}h - y_0)b; \qquad f = \tfrac{2}{3}P \div (\tfrac{1}{2}h - y_0)b.$$

If the case is that of a wall, and $P$ is the resultant force per unit of length, $b = 1$.

As $y_0$ increases, $f$ increases, until finally the stone crushes at the edge of the joint, or shears on an oblique plane as described in § 23. Sometimes the pressure is not well distributed, from poor bedding of the stones, and *spalls*, or chips, under the action of the shearing above referred to, may break off along the edge, without failure being imminent, since when the *high* spots break off others come into bearing.

$P$ can never traverse one edge of the joint, if tension is not possible at the other edge, as the unit stress then becomes infinite. Some writers commit an error in determining the thickness of a wall by equating the moment of the overturning force about the front edge or toe with the moment of the weight of the wall about the same point. This process is equivalent to making the action line of $P$ traverse that point. The centre of moments should be taken either at the outer edge of the middle third of the joint, when pressure is desired over the whole joint, or about a point at such a distance $\frac{1}{2}h - y_0$ from the front as will give maximum safe pressure at the front edge. A uniformly varying stress extending over three times that distance will equal $P$, as lately stated. A portion of the joint at the rear will then tend to open.

*Examples.*—A short, hollow, cylindrical column, 12 in. external diameter, 10 in. internal diameter, supports a beam which crosses the column 2 in. from its centre.

$$\frac{8d}{d^2 + d'^2} = \frac{8 \cdot 12}{144 + 100} = \frac{96}{244}. \qquad f = \frac{P}{S}\left(1 \pm \frac{2 \cdot 96}{244}\right) = f_c(1 \pm 0.8),$$

or the stress at either edge will be 80% greater and less than the mean stress.

A joint, 10 ft. broad, of a retaining-wall is cut at a point 3 ft. 9 in. from the front edge by the line of the resultant thrust above that joint. If this thrust per foot of length of the wall is 28,000 lb., the pressure per square foot at the front edge will be $\frac{28,000}{10}(1 + 1\frac{1}{4} \cdot \frac{6}{10}) = 1\frac{3}{4} \cdot 2,800 = 4,900$ lb. At the rear it will be 700 lb. per sq. ft.

**126. Resistance of Columns.**—A column, strut, or other piece, subjected to longitudinal pressure, is shortened by the compression. As perfect homogeneousness does not exist in any material, the longitudinal elements will yield in different amounts, so that there is apt to be a slight, an imperceptible tendency to curvature of the strut. Hence the action line of the applied load may not traverse the centres of all cross-sections of the piece. The product of the applied force into the perpendicular distance of its line of action from the centre of any cross-section will be a bending moment, which must develop a resisting moment at the cross-section, resulting in a varying stress, as in § 124. This curvature under longitudinal pressure can be readily obtained with test specimens of most materials, even with some samples of cast iron, and the form of the curve apparently conforms to the one to be deduced by theory. A *tendency* to such a curve must therefore exist under working stresses, although the curvature is imperceptible, unless the column happens to have its load perfectly axial, a contingence that cannot be safely relied upon. The column formula, so called, should therefore be confidently applied.

Further, as such curvature can be produced in test specimens not more than four or five diameters long, such a formula is applicable to columns and struts of any length. It is not necessarily to be applied, however, to very short posts, or blocks, for the relation $P = fS$ will determine their size with sufficient exactness.

**127. The Yield-point Marks the Column Strength.** — The influence of $y_0$ in determining the load a compression piece will carry has been shown in § 124 to be very marked. A column which has become sensibly bent under a load is very near complete failure. The moment of the load at the cross-section of greatest lateral deflection has then become so large that the stress on the extreme fibres passes the yield-point, and the great increase of deformation at and above the yield-point at once increases the bending moment greatly. Hence it is true that the yield-point marks nearly the ultimate compressive strength of materials when tested in column form.

Again, the fact that, in tests of large columns, a *very slight shifting* of the point of application of the load at either end has a decided influence on the amount of weight such a column will carry is a confirmation of the statements with which this discussion opened. It also has a bearing upon the truth of the theoretical deduction as to the effect of eccentric loading, as discussed in § 124, and to be applied to long columns later.

**128. Direction of Flexure.**—The flexure usually occurs, unless there is some defect or weakness, in a direction *parallel* to the *least transverse dimension* of the strut, *i.e.*, perpendicular to that axis in the cross-section which offers the least resisting moment. By the application of longitudinal pressure to a slender rod its flexure may be made very apparent. The *form* of the column formula ought to resemble that of § 124,

$$P = fS \div \left(1 + \frac{y_0 y_1}{r^2}\right).$$

**129. The Ideal Column. Euler's Formula.** — Assume the column to be perfectly straight and homogeneous and the load to be applied axially. Under such ideal conditions the load can cause no flexure, but if a small deflection is given the column by the application of a lateral force, it is desired to find what load is just sufficient to maintain that deflection after the lateral force is removed.

If the column is fixed in direction at its ends, by its connections to other pieces, or by having a broad, well-bedded base and cap, it will act in flexure much as a beam fixed at the ends. A couple or bending moment, which may be represented by $M_0$, will thus be introduced at each end. Let $P$ = applied external force or load; $v$ = any deflection ordinate, measured at right angles to the action line of $P$, from the original axis of the column to any point in the axis when bent; $x$ = distance from one end along the original axis to any ordinate $v$; $l$ = length of column.

Fig. 58

The combination of the moment $M_0$ at the end of the column with the force $P$ has the effect, as shown in mechanics, of shifting the force $P$ laterally a distance $M_0 \div P = v_0$; hence the action line of $P$ is now parallel to the original axis, at a distance $v_0$ from it, or in the line F E of Fig. 58. The ordinate to the points of contraflexure is therefore $v_0$. This action can be more fully realized by conceiving that the bearing surface at A is removed, and that $P$ acts at such a point on a horizontal lever as to keep the tangent to the curve at A strictly vertical.

As $v$ is measured from the original axis, the bending moment at any section is $M = P(v_0 - v)$, which will change sign when the second term is larger than the first.

## COLUMNS, POSTS, AND STRUTS.

If the flexure is very slight, an equation similar to that used with beams may be written

$$EI\frac{d^2v}{dx^2}=P(v_0-v).$$

Multiply by $\frac{dv}{dx}$ and integrate.

$$\frac{EI}{2}\left(\frac{dv}{dx}\right)^2=P(v_0v-\tfrac{1}{2}v^2)+C.$$

As $\frac{dv}{dx}=0$ when $v=0$, $C=0$. When $\frac{dv}{dx}=0$ at $D$, the middle of the length, $v_{max.}=2v_0$, or double the ordinate at the point of contraflexure, $F$. Extracting the square root of the above equation and separating the variables gives

$$dx=\sqrt{\frac{EI}{P}}\frac{dv}{\sqrt{2v_0v-v^2}};$$

$$x=\sqrt{\frac{EI}{P}}\text{ versin}^{-1}\frac{v}{v_0}+C'.$$

As $v=0$ when $x=0$, $C'=0$.

$$v=v_0\text{ versin}\left(x\sqrt{\frac{P}{EI}}\right)=v_0\left(1-\cos\left(x\sqrt{\frac{P}{EI}}\right)\right).$$

As $1-\cos\theta=2\sin^2\tfrac{1}{2}\theta$,

$$v=2v_0\sin^2\left(\tfrac{1}{2}x\sqrt{\frac{P}{EI}}\right).$$

If, in this equation, $x=\tfrac{1}{2}l$ a value of $v_{max.}$ is obtained to be equated with the previous value, $2v_0$.

$$v_{max.}=2v_0=2v_0\sin^2\left(\tfrac{1}{4}l\sqrt{\frac{P}{EI}}\right).$$

$$\sin\left(\tfrac{1}{4}l\sqrt{\frac{P}{EI}}\right)=1;\quad \tfrac{1}{4}l\sqrt{\frac{P}{EI}}=\tfrac{1}{2}\pi.$$

$$P=\frac{4\pi^2EI}{l^2},$$

which is commonly known as *Euler's formula*. $P$ is the ultimate strength of the column and is independent of the deflection; that is, within certain limits $P$ is the same, no matter what small deflection is given the column. Under a load less than $P$ the column would straighten; under a greater load it would fail.

The equation of the curve of the column is

$$v = 2v_0 \sin^2\left(\frac{\pi}{l}x\right).$$

To find the points of contraflexure, make $v = v_0$.

$$\sin\left(\frac{\pi}{l}x\right) = \sqrt{\tfrac{1}{2}} = \sin 45° = \sin \frac{\pi}{4}.$$

$$\therefore \frac{x}{l} = \tfrac{1}{4} \quad \text{or} \quad x = \tfrac{1}{4}l \text{ from either end.}$$

Hence the curve is made of four equal portions, A F, F D, D E, and E H.

A column hinged, pin-ended, or free to turn at its ends, and of length represented by $E F = \tfrac{1}{2}l$, will have the same portion of stress at D that is due to *bending* as does a column of length $A H = l$, which is fixed at its ends.

If, in actual cases, F is considered to be practically in the same position horizontally as before loading, it may be said that a column fixed at one end and hinged at the other, of length $F H = \tfrac{3}{4}l$, will also have the same portion of stress arising from bending. The maximum deflection will then occur at one-third of its length from the hinged end. This result has been verified by direct experiment on a full-sized steel bridge member.

**130. Rankine's Column Formula.**—Euler's formula can be put in the form

$$\frac{P}{S} = \frac{4\pi^2 E}{\left(\dfrac{l}{r}\right)^2}.$$

This curve, when plotted with $\dfrac{P}{S}$ and $\dfrac{l}{r}$ as variables, has the form of the curve B C, Fig. 59, where the greatest mean unit stress which any ideal column of a given ratio of $l$ to $r$ can endure is given by the ordinate to the curve. But as the yield-point marks the ultimate strength of metal in compression, for short columns the line A B, whose ordinate is the yield-point, takes the place of the curve.

For long, slender columns (when $l \div r$ exceeds 200, say) Euler's formula gives results close to the ultimate strength found by loading actual columns to destruction; but for shorter columns the formula gives results much too large. It must be remembered that actual columns do not satisfy the conditions from which Euler's formula was derived. No real column is perfectly straight

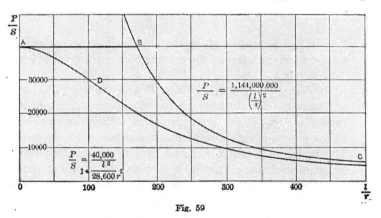

Fig. 59

and homogeneous, nor can the load be applied exactly along the axis, and as no two columns are exactly alike no theoretical formula can exactly fit all cases. For these reasons Euler's formula is never used in designing.

When the results of actual column tests are plotted, they follow in a general way some such curve as A D C. *Rankine's formula*, $\dfrac{P}{S} = \dfrac{p}{1 + a\dfrac{l^2}{r^2}}$, is the equation of such a curve, in which

$p$ is the yield-point of the material and $a$ is a numerical coefficient. By modifying the value of $a$ the curve can be made to agree fairly well with the results of tests. For a short block when

$l \div r$ approaches zero the formula becomes $P \div S = p$. If $a$ is made equal to $p \div 4\pi^2 E$, the curve of Rankine's formula approaches Euler's for large values of $l \div r$, where Euler's agrees well with the results of tests. This value of $a$ has been used by some authorities and is believed to give conservative values. The curve of Fig. 59 was so plotted. However, the value of any column formula depends upon its agreement with the results of tests, and in practice $a$ is generally an empirical factor.

In order that $P$ of Rankine's formula may be the load which an actual column can safely carry instead of its ultimate strength, the greatest allowable unit stress on the material, $f$, is to be substituted for the yield-point, $p$. The formula then becomes

$$\frac{P}{S} = \frac{f}{1 + a\frac{l^2}{r^2}}.$$

It is to be regarded as giving the *mean* unit stress on any cross-section when the stress on the extreme fibre at the point of greatest deflection is $f$.

In designing columns the load, $P$, and the length, $l$, are usually known. The form of cross-section is then assumed which fixes $S$ and $r$. If the assumed values satisfy the equation the section chosen is satisfactory; if not, another trial must be made. The radius of gyration, $r$, to be used is the one perpendicular to the neutral plane, that is, it is measured in the direction in which the column is likely to deflect. Unless the column is restrained in some way it will, of course, deflect in the direction of the least radius of gyration of the cross-section.

*Example.*—What load will a 12-in. $31\frac{1}{2}$-lb. I beam 10 ft. long carry as a column if the mean unit stress is $12{,}000 \div \left(1 + \frac{l^2}{36{,}000 r^2}\right)$? $r$ about axis through web $= 1.01$ in. $S = 9.26$ sq. in. $\frac{l^2}{r^2} = \frac{14{,}400}{1.02} = 14{,}100$. $\frac{14{,}100}{36{,}000} = 0.392$. $\frac{12{,}000}{1.392} = 8{,}620$ lb. $P = 8{,}620 \times 9.26 = 79{,}800$ lb.

**131. Multipliers of a.**—As seen above, $2l$ should theoretically be substituted for $l$ when the column is hinged or free to turn at its ends, in order to obtain the equivalent length of a column which is fixed at the ends; and for a column fixed at one end and hinged at the other $\frac{4}{7}l$ should be substituted for $l$ for the same reason. Or, more conveniently, $a$ may be used for a column fixed at the ends and of length $l$; $4a$ for a column hinged at both ends and of length $l$; and $\frac{16}{9}a$ for a column hinged at one end and fixed at the other, length $l$. Actual tests, carried, however, to the extreme of bending or crippling, appear to show that a column bearing on a pin at each end is not hinged or perfectly free to turn; hence the multipliers of $a$ more commonly used, instead of 1, $\frac{16}{9}$, and 4, are 1, $\frac{3}{2}$, and 2.

**132. Pin Friction.**—Some regard columns as neither perfectly fixed nor perfectly hinged, and use but one value of $a$ for all, which might perhaps then be taken as a mean value. The moment of friction on a pin is considerable. If $P$ is the load on a post or strut, $d$ the diameter of the pin, and $\tan \phi$ the coefficient of friction of the post on the pin, the moment of friction at the pin will be $P \cdot \frac{1}{2}d \cdot \tan \phi$; and this moment, if greater than $M_0 = Pv_0$, will keep the post restrained at the end, so that the tangent there to the curve remains in its original direction. As $P$ and $v_0$ increase, $M_0$ will become the greater when $v_0$ exceeds $\frac{1}{2}d \tan \phi$; the column will then be imperfectly restrained at its ends, and the inclination will change. As the friction of motion is less than that of rest, such movement when started may be rapid. Some tested columns, showing at first the curve of Fig. 58 known as triple flexure, have suddenly sprung into a single curve and at once offered less resistance.

It may be doubted whether ordinary columns under working loads ever develop a value of $v_0$ sufficient to overcome the pin friction, unless the column is very slender and the diameter of the pin small. The custom is quite general of using the same column formula for struts with fixed and with hinged ends.

**133. Straight-line Column Formula.**—In engineering structures columns which have a greater ratio of length to radius of gyration than 100 or 150 are very rarely used. Within this limit a straight line can be drawn which will represent the average results of a series of column tests, plotted as described in § 130, as well as Rankine's formula does. As the equation of a straight line is easier to solve than the equation of a curve, the *straight-line* formula is much used. For working values it has the form

$$\frac{P}{S} = f - c\frac{l}{r},$$

in which $f$ is the greatest allowable unit stress on the material and $c$ is a numerical coefficient determined empirically. In this country both Rankine's and the straight-line formulas are extensively used in designing columns. Examples of each will be found in Chapter X.

**134. Swelled Columns.**—Some posts and struts, especially such as are built up of angles connected with lacing, are swelled or made of greater depth in the middle. If the strut is perfectly free to turn at the ends, such increase in the value of $r^2$ may be quite effective, and $r^2$ for the middle section may be used in determining the value of $P$, provided the latter does not too closely approach the uniform compression value at the narrower ends. But if the strut is fixed at the ends, or is attached by a pin whose diameter is large enough to make a considerable friction-moment, such enlargement at the middle is useless; for an equally large value of $r^2$ ought to be found at the ends also. Hence swelled columns and struts are but little used.

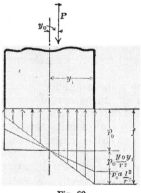

Fig. 60

**135. Column Eccentrically Loaded.**—When the load is applied eccentrically to a long column, the maximum unit stress found in the extreme fibre on the concave side must be due to three combined effects:

1st. The stress due to the load $P$, or $p_0 = P \div S$. (Fig. 60.)

2d. The stress due to the resisting moment set up by $P y_0$.

3d. The stress due to the resisting moment set up by $P v_0$.

From § 124, $\quad f = \dfrac{P}{S}\left(1 + \dfrac{y_0 y_1}{r^2}\right) = p_0 + p_0 \dfrac{y_0 y_1}{r^2}.$

From § 130, $\quad f = \dfrac{P}{S}\left(1 + a\dfrac{l^2}{r^2}\right) = p_0 + p_0 a\dfrac{l^2}{r^2}.$

If then the column is long and the line of action of $P$ deviates from the original axis of the column by a distance $y_0$, the three expressions, $p_0$, $p_0 y_0 y_1 \div r^2$, and $p_0 a l^2 \div r^2$ should be added, giving

$$f = p_0 \left(1 + a\frac{l^2}{r^2} + \frac{y_0 y_1}{r^2}\right). \quad \therefore \frac{P}{S} = \frac{f}{1 + a\frac{l^2}{r^2} + \frac{y_0 y_1}{r^2}}.$$

This formula may be put in a form more convenient for designing:

$$S = \frac{P}{\dfrac{f}{1 + a\dfrac{l^2}{r^2}}} + \frac{P y_0 y_1}{f r^2}.$$

The first term gives the area of cross-section necessary to resist the direct thrust and the second term the area to resist the bending moment due to eccentric loading.

Since $y_0$ will determine the direction of flexure, $r$ must be taken in this case in the direction $y_0$; that is, the moment of inertia and $r$ must be obtained about the axis through the centre of gravity and lying in the plane of the section, perpendicular to $y_0$. That the moment $P y_0$, although small, has a decided weakening effect on a column is proved by experiment, and its unintended presence may explain some anomalies in tests.

**136. Transverse Force on a Column.**—The resisting power of a column or strut to which a transverse force is applied in addition to the load in the direction of the axis, and the proper dimensions of the strut, are involved in some doubt. Theoretically, the formula is deduced as follows:

From the formula for the resisting moment of a beam, $M = fI \div y_1$, the stress on the outer layer from such bending moment is $My_1 \div I$. Hence if $M$ is *that particular value* of the bending moment (from the transverse load or force) which exists at the section of maximum strut deflection, *where the column stress is greatest* (that is, at the middle for a column with hinged ends, but perhaps at the ends for a column with fixed ends, since $M$ may then be greater at the ends), the maximum unit stress on the concave side

$$f = \frac{P}{S}\left(1 + a\frac{l^2}{r^2}\right) + \frac{My_1}{I}.$$

$$\frac{P}{S} = \frac{\left(f - \frac{My_1}{I}\right)}{1 + a\frac{l^2}{r^2}}.$$

$$S = \frac{P}{\dfrac{f}{1 + a\dfrac{l^2}{r^2}}} + \frac{My_1}{fr^2}.$$

If the straight-line formula is used instead of Rankine's, the last equation has the form

$$S = \frac{P}{f - c\dfrac{l}{r}} + \frac{My_1}{fr^2}.$$

Or, again, it may be said that at the section in question $P$ is moved laterally by the moment $M$ a distance $y_0 = M \div P$. Then, by § 135,

$$\frac{P}{S} = \frac{f}{1 + a\dfrac{l^2}{r^2} + \dfrac{My_1}{Pr^2}},$$

which reduces to the forms given above.

The value of $r^2$ to be used in these formulas is that for flexure in the plane of $M$.

*Example.*—A deck railway-bridge with 20-ft. panels has ties laid directly on the top chord, thus imposing a load of 2,500 lb. per foot. If the direct thrust is 249,000 lb., what should be the size of chord for a working stress on columns of $10,000 - 45l \div r$ and on beams of 9,000 lb. per sq. in.? Try a section composed of two 20-in. 65-lb. I beams and one $24 \times \frac{1}{2}$-in. cover-plate. $S = 50.16$ sq. in. The centre of gravity of the section lies 2.45 in. above middle of I beams. $y_1$ to upper fibre $= 8.05$ in. $I = 3,298$ in.$^4$ $r$ about horizontal axis $= 8.11$ in. $M = \frac{1}{8} \cdot 2,500 \cdot 20 \cdot 20 \cdot 12 = 1,500,000$-in.-lb. Working column stress $= 10,000 - 45 \times 240 \div 8.11 = 8,670$ lb. per sq. in. $249,000 \div 8,670 = 28.7$ sq. in. required for column action. $\dfrac{1,500,000 \times 8.05}{9,000 \times 8.11 \times 8.11} = 20.4$ sq. in. required for beam action. $28.7 + 20.4 = 49.1$ sq. in., total. Assumed section is sufficient.

**137. Lacing-bars.**—The parts of built-up posts are usually connected with lacing straps or bars. These bars carry the shear due to the bending moment arising from the tendency of the post to bend and should be able to stand the tension and compression induced by the shear. At the ends and where other members are connected, in order to insure a distribution of load over both members, batten or connection plates at least as long as the transverse distance between rivet rows are used in good practice.

Each piece should be of equal strength throughout all its details. A post or strut composed of two channels connected by lacing-bars and tie-plates is proportioned for a certain load, the mean unit stress being reduced in accordance with the formula in which the variable is the ratio of the length to the least radius of gyration of the whole section. In the lengths between the lacing-bars, this ratio for one channel with its own least radius should not be greater than for the entire post. Nor should the flange of the channel have any greater tendency to buckle than should one channel by itself. The same thing applies to the ends of posts, where flanges are sometimes cut away to admit other members. Quite a large bending moment may be thrown on such ends, when the plane of a lateral system of bracing does not pass through the pins or points of connection of the main truss system.

The usual bridge specification reads: Single lacing-bars shall have a thickness of not less than $\frac{1}{40}$, and double bars, connected by a rivet at their intersection, of not less than $\frac{1}{60}$ of the distance between the rivets connecting them to the member. Their width shall be: For 15-in. channels, or built sections with $3\frac{1}{2}$- and 4-in. angles, $2\frac{1}{2}$ in., with $\frac{7}{8}$-in. rivets; for 12- or 10-in. channels, or built sections with 3-in. angles, $2\frac{1}{4}$-in., with $\frac{3}{4}$-in. rivets; for 9- or 8-in. channels, or built sections with $2\frac{1}{2}$-in. angles, 2 in., with $\frac{5}{8}$-in. rivets.

All segments of compression members, connected by lacing only, shall have ties or batten plates placed as near the ends as practicable. These plates shall have a length of not less than the greatest depth or width of the member and shall have a thickness of not less than $\frac{1}{50}$ of the distance between the rivets connecting them to the compression member.

*Examples.*—1. A single angle-iron $6 \times 4 \times \frac{3}{8}$ in. and 6 ft. 8 in. long is in compression. Use $r = 0.9$ or obtain it from § 76. $S = 3.61$ sq. in. If $P \div S = 12{,}000 - 34l \div r$, what will it carry? 32,400 lb.

2. A square timber post 16 ft. long is expected to support 80,000 lb.

If $f = 1,000$ and the subtractive term is $2l \div r$, what is the size?
$10 \times 10$ in.

3. What is the safe load on a hollow cylindrical cast-iron column 13 ft. 6 in. long, 6 in. external diameter, and 1 in. thick, if it has a broad, flat base, but is not restrained at its upper end? ($f = 9,000$ lb., $E = 17,000,000$, $S = 15.7$ sq. in.) 124,000 lb.

4. If a short wooden post 12 in. square carries 28,800 lb. load, and the centre of pressure is 3 in. perpendicularly from the middle of one edge, what will be the maximum and the mean unit pressure, and the maximum unit tension, if any? 500 lb.; 200 lb.; $-100$ lb.

## CHAPTER X.

### SAFE WORKING STRESSES.

**138. Endurance of Metals Under Stress.**—It is important to determine how long a piece may be expected to endure stress when constant, when repeated, or when varied. and perhaps reversed; and it is still more important to find what working stresses may be allowed upon a given material in order that rupture by the stresses may be postponed indefinitely.

The experiments carried on by Wöhler and Spangenberg, and afterwards continued by Bauschinger, show the action of iron and steel under repeated stress. Tests were made on specimens in tension, bending, and torsion.* A number of bars of wrought iron and steel were subjected repeatedly to a load which varied between zero and an amount somewhat less than the ultimate strength. When the bar broke under the load a similar bar was tested under a reduced load and was found to resist a greater number of applications before rupture. Finally a load was reached which did not cause rupture after many million repetitions. The stress caused by such load was taken as the safe strength of the material and was called the *primitive safe strength* The following table is an illustration:

WÖHLER'S TESTS OF PHŒNIX IRON IN TENSION.

| Stress varying between | Number of Applications. |
|---|---|
| 0 and 46,500 lb. per sq. in. | 800 |
| 0 " 43,000 " " " " | 107,000 |
| 0 " 39,000 " " " " | 341,000 |
| 0 " 35,000 " " " " | 481,000 |
| 0 " 31,000 " " " " | 10,142,000 |

* For a description of machines and tests, see Unwin, Testing of Materials of Construction.

As a result of Wöhler's and Spangenberg's experiments the latter states that alternations of stress may take place between the following limits of tensile (−) and compressive (+) stress in pounds per square inch with equal security against rupture:

### PHŒNIX IRON AXLE.

| | | | |
|---|---|---|---|
| −15,500 | +15,500 | Difference | 31,000 |
| −29,000 | 0 | " | 29,000 |
| −43,000 | −23,500 | " | 19,500 |

### KRUPP CAST-STEEL AXLE.

| | | | |
|---|---|---|---|
| −27,000 | +27,000 | Difference | 54,000 |
| −47,000 | 0 | " | 47,000 |
| −78,000 | −34,000 | " | 44,000 |

### UNHARDENED SPRING STEEL.

| | | | |
|---|---|---|---|
| −48,500 | 0 | Difference | 48,500 |
| −68,000 | −24,000 | " | 44,000 |
| −78,000 | −39,000 | " | 39,000 |
| −87,500 | −58,500 | " | 29,000 |

The figures are approximate.

The results illustrate Wöhler's law, that *rupture of material may be caused by repeated alternations of stress none of which attains the absolute breaking limit. The greater the range of stress the smaller the stress which will cause rupture.*

The term *fatigue of metals* is often applied to the phenomena just described.

**139. Bauschinger's Experiments.** — Professor Bauschinger, besides making tests similar to Wöhler's, determined the effect which repeated stresses had upon the position of the elastic limit and the yield-point. (§ 13, 14, and 15.) He derived the following results from tensile tests:

| | Primitive Safe Stress, $u$. | Original Elastic Limit. | Original Yield-point. | Ultimate Strength, $t$. |
|---|---|---|---|---|
| Wrought-iron plate | 28,000 | 14,800 | 29,700 | 54,600 |
| Bessemer soft-steel plate | 34,000 | 33,900 | 41,800 | 62,000 |
| Iron, flat | 31,000 | 25,700 | 32,600 | 57,600 |
| Iron, flat | 34,000 | 32,300 | 35,100 | 57,200 |
| Thomas steel axle | 43,000 | 38,100 | ..... | 87,000 |

From an inspection of the table it is seen that the primitive safe stress on sound bars of iron and steel for tensions alternating from zero to that stress is a little less than the original yield-point of the metal. Although the safe stress was found to be greater than the original elastic limit, yet when bars which remained unbroken after several million applications of stress were tested statically it was found that the elastic limit had risen above the stress applied. The inference is that the primitive safe stress is below the elastic limit although the elastic limit may be an artificially raised one.

Pulling a bar with loads above the original elastic limit and below the yield-point raises the elastic limit in tension but lowers it in compression and vice versa. These artificially produced elastic limits are very unstable. When a bar is subjected to a few alternations of equal and opposite stresses which are equal to or somewhat exceed the original elastic limits, the latter tend toward fixed positions called by Bauschinger *natural elastic limits*. He advanced the view that the original elastic limits of rolled steel and iron are artificially raised by the stresses set up during manufacture. These natural elastic limits seem to correspond with Wöhler's range of stress for unlimited alternating stresses.

**140. Alteration of Structure.**—If a bar has been subjected to repetitions of stress somewhat below the yield-point, any general weakening or deterioration in the quality of the bar ought to show itself in some way; but bars which have endured millions of repeated stresses or bars which have broken under repeated stresses give no indication of any weakening when tested statically. Instead of being weakened by repeated stresses well below the primitive safe strength, a bar is really improved in tenacity.

Bars fractured in the Wöhler machines did not draw out, no matter how ductile the material might be, but broke as if the material were brittle. From these facts we are forced to conclude that whatever weakness there may be is confined to the surface of rupture. Such breaks appear to be detail breaks. Continuity is first destroyed at a flaw or overstressed spot, and from that point the fracture spreads gradually until the section is so weakened that the load is eccentric and bending stresses are set up which cause the piece to break by flexure.

From the results of Wöhler's and Bauschinger's tests, it would appear that steel might undergo an indefinite number of repetitions of stress within the elastic limit without rupture, or that within that limit the elasticity was perfect. However, certain experiments on the thermoelectric and magnetic properties of iron under stress and on the molecular friction of torsional pendulums give results inconsistent with the theory of perfect elasticity. Engineers differ more or less in their interpretations of the experimental results but agree in using lower working stresses for varying than for static loads.

**141. Launhardt-Weyrauch Formula.**—A number of formulas based on Wöhler's experiments have been advanced for the determination of the allowable unit stress on iron and steel when the range, as well as the magnitude, of the stress is considered. Launhardt proposed the following formula for the breaking load of a member which is subjected to stresses varying between a maximum and a minimum stress of the same kind:

$$a = u\left(1 + \frac{t-u}{u}\frac{\text{min. stress}}{\text{max. stress}}\right).$$

$a$ = breaking load under the given conditions.
$u$ = primitive safe stress.
$t$ = ultimate strength, static.
$v$ = greatest stress piece will bear if repeated from $+v$ to $-v$ indefinitely.

Weyrauch extended Launhardt's formula to cover cases of alternate tension and compression, in which case the minimum stress is to be considered as negative.

$$a = u\left(1 + \frac{u-v}{u}\frac{\text{min. stress}}{\text{max. stress}}\right).$$

As the results of the different experiments vary more or less, the different authorities do not agree as to the values of $t$, $u$, and $v$ to be substituted in the formulas. In some of Wöhler's experiments $u$ appears to be greater than $\frac{1}{2}t$ and to approximate to $\frac{3}{4}t$. This value is assumed by Weyrauch and gives $(t-u) \div u = \frac{1}{3}$. Likewise if $v = \frac{1}{2}u$ the coefficient of the second formula becomes $\frac{1}{2}$ and both formulas reduce to

$$a = u\left(1 + \tfrac{1}{2}\frac{\text{min. stress}}{\text{max. stress}}\right),$$

the sign of the second term changing for reversed stresses. The choice of values of $u$ and $v$ to use in the equations seems to have been determined largely by the result sought. The formulas are of rather doubtful value, but they are more or less used for determining working stresses in steel bridges.

The following relationships of $t$, $u$, and $v$ are given by some authorities, and they conform more nearly with the results of the experiments quoted in this chapter:

Steady load.......................... $a = t$
Load varying from 0 to $u$............. $a = u = \tfrac{1}{2}t$
    "      "      " $+v$ to $-v$............. $a = v = \tfrac{1}{3}t$

**142. Reduction of Unit Stresses.**—The safe working stress to be used for any material will depend upon several considerations: Whether the structure is to be temporary or permanent; whether the load is stationary or variable and moving; if moving, whether its application is accompanied by severe dynamic shock and perhaps pounding; whether the load assumed for calculation is the absolute maximum; whether such maximum is applied rarely or is likely to occur frequently; whether the stresses obtained are exact or approximate; whether there are likely to be secondary stresses due to moments arising from changes of the assumed frame; what the importance of the piece is in the structure, and the possible damage that might be caused by its failure.

The allowable unit stresses of different kinds, and for greater or less change of load, will be further reduced to provide against: Distribution of stress on any cross-section somewhat different from that assumed; variations in quality of material; imperfections of workmanship, causing unequal distribution of stress; scantness of dimensions; corrosion, wear, or other deterioration from lapse of time or neglect; lack of exactness of calculation.

The allowable unit stresses so determined will be but a small fraction of the ultimate or breaking strength of the material; and it is evident that the idea that it will require several times the allowable maximum working load to cause a structure to fail is seriously in error.

Overconfidence of the inexperienced designer in the correctness of his design may be checked by a study of this section.

**143. Load and Impact.**—The design should be completely carried out, both in the principal parts and in the details. The latter require the most careful study, that they may be at once effective, simple, and practical.* All the exterior forces which may possibly act upon a structure should be considered, and due provision should be made for resisting them. The static load, the live load, pressure from wind and snow, vibration, shock, and centrifugal force should be provided for, as should also deterioration from time, neglect, or probable abuse. A truss over a machine-shop may at some time be used for supporting a heavy weight by block and tackle, or a line of shafting may be added; a small surplus of material in the original design will then prove of value. Light, slender members in a bridge truss, while theoretically able to resist the load shown by the strain sheet, are of small value in time of accident. The tendency from year to year is towards heavier construction.

Secondary stresses, as they are called, are due—first, to the moments at intersections or joints, when the axes of the members coming together at a connection do not intersect at a common point; and second, to the moments set up at joints by the resistance to rotation experienced by the several parts when the frame or truss is deflected by a moving load. If symmetrical sections are used for the members, if the connecting rivets are symmetrically placed, and if the axes of the intersecting members meet at one point, secondary stresses will be much reduced.

All members of a structure should be of equal strength, and the connections should develop the full strength of the body of the members connected. The connections should be as direct as possible. When a live load is joined to a static load, the judgment of the designer, or of the one who prepares the specifications for the designer, must be exercised. A warehouse floor to be loaded with a certain class of goods has maximum stresses from a static load. The floor of a drill-room, ball-room, or highway bridge receives maximum loading from a crowd of people

---

* A portion of these paragraphs is extracted from a lecture by Mr. C. C. Schneider.

the possible density of which can be ascertained. But if these masses of people keep moving, and more particularly if they keep step, the effect of their weight will be increased by the vibrations resulting therefrom. This action is generally called *impact*.

In the case of a building, the floor-joists, receiving the impact directly, will be most affected; the girders which carry the joists will be less affected, and the columns which support the girders will receive a smaller percentage of the impact, the proportionate effect growing less as the number of stories below the given floor increases. In the absence of trustworthy data from which to determine this impact, it is left to the judgment of the engineer to increase the live load by a certain percentage, or to decrease the allowable unit stress, for each case, to provide for the effect, as will be seen in the values given later.

For economy, make designs which will simplify the shop work, reduce the cost, and insure ease of fitting and erection. Avoid an excess of blacksmith work and much use of bent pieces.

**144. Dead and Live Load.**—From what has been said it is readily seen that a moving or live load has a much more serious effect on a truss than a static or dead load, and in the case of a railroad bridge the stresses in the members due to a rapidly moving train are much greater than the stresses would be under such a train at rest. In designing the members of steel bridges this increase of stress is usually provided for by one of three ways:

Using different working stresses for dead and live load, as a unit stress of 18,000 lb. per sq. in. for dead load and 9,000 lb. for live. When this method is employed the ratio is commonly 2 to 1.

Using a varying unit stress found from a modified form of the Launhardt-Weyrauch formula, as $f = 9{,}000 \left(1 + \dfrac{\text{min. stress}}{\text{max. stress}}\right)$. If the stresses reverse the fraction is negative. The coefficient $\frac{1}{2}$ in the original formula is dropped arbitrarily.

Adding a certain percentage of the live load, called *impact*, to the sum of the dead- and live-load stresses, and using a constant working stress as 18,000 lb., one impact formula often used is

$$\text{Impact} = \frac{300}{300 + \text{length}} \times \text{live-load stress}.$$

By "length" is meant the distance in feet through which the load must pass to produce maximum stress in the member in question. Such experiments as have been made on bridges under moving load indicate that the impact is not as great as the formula gives. The third method is sometimes combined with the first or second, but with the use of a different impact formula.

While these three ways of designing are quite different in theory and method, the resulting sizes of members are much the same.

**145. Working Stresses for Timber.**—The following table is from Bulletin No. 12, U. S. Dept. of Agriculture, Div. of Forestry:

SAFE UNIT STRESSES AT 18% MOISTURE.
Structures freely exposed to the weather.

|  | Compression. | | Bending. | Shear. | Tension. |
|---|---|---|---|---|---|
|  | With Grain. | Across Grain. | Extreme Fibre. | With Grain. |  |
| Long-leaf pine....... | 1,000 | 215 | 1,550 | 125 | 12,000 |
| Short-leaf pine...... | 840 | 215 | 1,300 | 100 | 9,000 |
| White pine.......... | 700 | 150 | 880 | 75 | 7,000 |
| Norway pine........ | 760 | 140 | 1,090 | | |
| Douglas spruce...... | 880 | 170 | 1,320 | | |
| Redwood........... | 650 | 115 | | | |
| White oak.......... | 800 | 400 | 1,200 | 200 | 10,000 |
| Factor of safety..... | 5 | 3 | 5 | 4 | 1 |

The values given were found by taking the average of the lowest 10 per cent. of the results of tests on non-defective timber and dividing by the factor of safety given in the last row. As timber never fails in tension the safe tensile stress is not given; the last column shows the ultimate strength. For structures protected from the weather the compressive and bending values may be increased 20 per cent. The shearing value should not be increased. The table was constructed for designing railway trestles.

The following safe working stresses have been recommended by the Committee on Strength of Bridge and Trestle Timbers, Am. Assn. of Ry. Supts. of Bridges and Buildings:

## SAFE WORKING STRESSES.

|  | Tension. | | Compression. | | | Bending. | Shear. | |
|---|---|---|---|---|---|---|---|---|
|  | | | With Grain. | | | | | |
|  | With Grain. | Across Grain. | End Bearing. | Cols. under 15 Diams. | Across Grain. | Extreme Fibre. | With Grain. | Across Grain. |
| Long-leaf pine....... | 1,200 | 60 | 1,600 | 1,000 | 350 | 1200 | 150 | 1,250 |
| White pine.......... | 700 | 50 | 1,100 | 700 | 200 | 700 | 100 | 500 |
| Norway pine......... | 800 | ..... | 1,200 | 800 | 200 | 700 | | |
| Douglas spruce....... | 1,200 | ..... | 1,600 | 1,200 | 300 | 1,100 | 150 | |
| Spruce and Eastern fir. | 800 | 50 | 1,200 | 800 | 200 | 700 | 100 | 750 |
| Hemlock............ | 600 | ..... | ..... | 800 | 150 | 600 | 100 | 600 |
| Redwood............ | 700 | ..... | ..... | 800 | 200 | 750 | 100 | |
| White oak........... | 1,000 | 200 | 1,400 | 900 | 500 | 1,000 | 200 | 1,000 |
| Factor of safety...... | 10 | 10 | 5 | 5 | 4 | 6 | 4 | 4 |

Messrs. Kidwell and Moore deduced the following column formulas based on the values given in the preceding table. Least cross-dimension of column $=h$; length $=l$.

|  | $l \div h \leq 25$ | $l \div h > 25$ |
|---|---|---|
| Long-leaf pine.......... | $1{,}000 - 8.5\,l \div h$ | $1{,}000 - 11\,l \div h$ |
| White pine............. | $700 - 6.\ l \div h$ | $700 - 8\,l \div h$ |
| Norway pine, spruce, and Eastern fir.......... | $800 - 6.5\,l \div h$ | $800 - 8\,l \div h$ |
| Douglas spruce......... | $1{,}200 - 9.\ l \div h$ | $1{,}200 - 12\,l \div h$ |
| Hemlock............... | $800 - 7.6\,l \div h$ | $800 - 9.5\,l \div h$ |
| Redwood............... | $800 - 8.6\,l \div h$ | $800 - 11\,l \div h$ |
| Oak.................... | $900 - 9.\ l \div h$ | $900 - 11\,l \div h$ |

For buildings Mr. C. C. Schneider recommends the following unit stresses:

|  | Compression. | | | Bending. | Shear with Grain. |
|---|---|---|---|---|---|
|  | With Grain. | | Across Grain. | Extreme Fibre. | |
|  | End Bearing. | Columns under 10 Diameters. | | | |
| Long-leaf pine...... | 1,500 | 1,000 | 350 | 1,500 | 100 |
| White pine, spruce,.. | 1,000 | 600 | 200 | 1,000 | 100 |
| Hemlock........... | 800 | 500 | 200 | 800 | 100 |
| White oak.......... | 1,200 | 1,000 | 500 | 1,200 | 200 |

For columns when $l \div h > 10$, $\dfrac{P}{S} = f - \dfrac{f}{100}\dfrac{l}{h}$, in which $f$ is the compressive stress for a short column taken from the table.

The strength of columns built up by bolting two or more pieces of timber together is no greater than that of the individual pieces acting independently.

The names of woods vary in different localities; long-leaf, short-leaf, and one or two other Southern species of pine are not distinguished in the market and are called Southern, Georgia, or yellow pine. Douglas spruce is also known as Oregon fir, spruce, or pine.

**146. Working Stresses for Railroad Bridges.**—Mr. Theodore Cooper's Specifications, edition of 1901, give the following unit stresses in pounds per square inch.

### MEDIUM STEEL IN TENSION.

| | |
|---|---|
| Main truss members—dead load................... | 20,000 |
| "      "      "      live  "   ..................... | 10,000 |
| Floor-beams and stringers.............................. | 10,000 |
| Lateral and sway bracing—wind load................. | 18,000 |
| "    "    "    "    live "   ................. | 12,000 |
| Floor-beam hangers and members liable to sudden loading | 6,000 |

For *soft steel* in tension reduce the unit stresses 10 per cent.

### MEDIUM STEEL IN COMPRESSION.

| | |
|---|---|
| Chord segments—dead load.................... | $20,000 - 90l \div r$ |
| "      "      live  "  .................... | $10,000 - 45l \div r$ |
| Posts of through-bridges—dead load........... | $17,000 - 90l \div r$ |
| "      "      "      live  "   ........... | $8,500 - 45l \div r$ |
| Posts of deck-bridges and trestles—dead load.... | $18,000 - 80l \div r$ |
| "    "    "    "    "    live "  .... | $9,000 - 40l \div r$ |
| Lateral and sway bracing—wind load........... | $13,000 - 60l \div r$ |
| "    "    "    "    live "  ........... | $8,700 - 40l \div r$ |

For *soft steel* in compression reduce the unit stresses 15 per cent.

The ratio of $l$ to $r$ shall not exceed 100 for main members and 120 for laterals.

When a member is subjected to alternating stresses, each stress shall be increased by eight-tenths of the lesser and the member designed for that stress which gives the larger section.

## SAFE WORKING STRESSES.

### PINS AND RIVETS.

Shear.................................... 9,000
Bearing (thickness × diameter)............ 15,000
Bending (pins)............................ 18,000
For floor rivets increase number of rivets 25%.
" field  "        "        "   "   "   50%.
" lateral "  decrease      "   "   "   33%.

### PEDESTALS.

Pressure on rollers, pounds per lineal inch, 300 × diam. in inches.

Pressure of bedplates on masonry, pounds per square inch, 250.

The specifications of the American Bridge Co., 1900, give the following working stresses:

The maximum stress in a member is found by adding the dead load, live load, and impact. Impact $=\dfrac{300}{300+L}\times$ live-load stress, in which $L$ is length in feet of distance to be loaded to produce maximum stress in member.

|  | Soft Steel. | Medium Steel. |
|---|---|---|
| Tension................. | 15,000 | 17,000 |
| Compression............. | $\dfrac{15,000}{1+\dfrac{l^2}{13,500r^2}}$ | $\dfrac{17,000}{1+\dfrac{l^2}{11,000r^2}}$ |
| Shear on web-plates..... | 9,000 | 10,000 |

The ratio of $l$ to $r$ shall not exceed 100 for main compression members and 120 for laterals.

When a member is subjected to alternating stresses the total section shall be equal to the sum of the areas required for each stress.

### PINS AND RIVETS.

Shear.............................. 11,000   12,000
Bearing (thickness × diameter)..... 22,000   24,000
Bending (pins)..................... 22,000   25,000
For hand-driven field rivets increase number 25%.
" power-driven  "      "       "        "    10%.

### PEDESTALS.

Pressure on rollers, pounds per lineal inch,
$$12,000\sqrt{\text{diam. in inches}}.$$

Pressure of bedplates on masonry, pounds per square inch, 400.

**147. Working Stresses for Highway Bridges.**—Mr. Cooper's Specifications, 1901, give the following stresses:

#### MEDIUM STEEL IN TENSION.

| | |
|---|---|
| Main truss members—dead load | 25,000 |
| "        "         "    live  " | 12,500 |
| Floor-beams, stringers, and riveted girders | 13,000 |
| Lateral and sway bracing—wind and live load | 18,000 |
| Floor-beam hangers and members liable to sudden loading | 8,000 |

For *soft steel* in tension reduce the unit stresses 10 per cent.

#### MEDIUM STEEL IN COMPRESSION.

| | |
|---|---|
| Chord segments—dead load | $24,000 - 110l \div r$ |
| "        "     live  " | $12,500 - 55l \div r$ |
| Posts of through-bridges—dead load | $20,000 - 90l \div r$ |
| "      "       "         live  " | $10,000 - 45l \div r$ |
| Posts of deck-bridges—dead load | $22,000 - 80l \div r$ |
| "     "       "       live  " | $11,000 - 40l \div r$ |
| Lateral and sway bracing—wind load | $13,000 - 60l \div r$ |
| "      "      "       "    live  " | $8,700 - 40l \div r$ |

For *soft steel* in compression reduce the unit stresses 15 per cent.

The ratio of $l$ to $r$ shall not exceed 100 for main members and 120 for laterals.

When a member is subjected to alternating stresses, each stress shall be increased by eight-tenths of the lesser and the member designed for that stress which gives the larger section.

#### PINS AND RIVETS.

| | |
|---|---|
| Shear | 10,000 |
| Bearing (diameter × thickness) | 18,000 |
| Bending (pins) | 20,000 |

For floor   rivets increase number 25%.
" field      "      "       "    50%.
" lateral    "    decrease  "    30%.

### PEDESTALS.

Pressure on rollers, pounds per lineal inch, 300 × diam. in inches.

Pressure of bed-plates on masonry, pounds per square inch, 250.

### TIMBER FLOOR-JOISTS.

Fibre stress on yellow pine or white oak.......... 1,200
" " " white pine or spruce............ 1,000

The working stresses given by the American Bridge Co. Specifications for Highway Bridges, 1901, are the same as for Railway Bridges, with the following exceptions:

The sum of dead and live load is to be increased by 25 per cent. of the live load to compensate for the effect of impact and vibration.

The ratio of $l$ to $r$ shall not exceed 120 for main members and 140 for laterals.

When a member is subjected to alternating stresses, design for that stress which gives the larger section.

Fibre stresses on floor-joists same as in preceding specification.

**148. Working Stresses for Buildings.**—A set of building specifications by Mr. C. C. Schneider may be found in the Transactions of the American Society of Civil Engineers, Vol. LIV, June, 1905. Some of the working stresses recommended are abstracted below.

Structural steel of ultimate strength between 55,000 and 65,000 lb. per sq. in. is to be used.

Tension............................ 16,000
Compression....................... $16,000 - 70l \div r$
Shear on web plates (gross section)..... 10,000

For wind bracing and the combined stresses due to wind and the other loading the permissible stresses may be increased 25 per cent.

The ratio of $l$ to $r$ shall not exceed 125 for main compression members and 150 for wind bracing.

Members subjected to alternating stresses shall be proportioned for the stress giving the larger section.

### PINS AND RIVETS.

| | |
|---|---|
| Shear on pins and rivets | 12,000 |
| "     "  bolts | 9,000 |
| Bearing on pins and rivets | 24,000 |
| "     "  bolts | 18,000 |
| Bending on pins | 24,000 |

For field rivets increase number 33 per cent.

### WALL PLATES AND PEDESTALS.

Bearing pressures in pounds per square inch on

| | |
|---|---|
| Brickwork, cement mortar | 200 |
| Rubble masonry, cement mortar | 200 |
| Portland cement concrete | 350 |
| First-class masonry, sandstone | 400 |
| "     "     "     granite | 600 |

### MASONRY.

Permissible pressure in masonry, tons per square foot:

| | |
|---|---|
| Common brick, natural cement mortar | 10 |
| "     "  Portland cement mortar | 12 |
| Hard-burned brick, Portland cement mortar | 15 |
| Rubble masonry, Portland cement mortar | 10 |
| Coursed rubble     "     "     " | 12 |
| Portland cement concrete, 1-2-5 | 20 |

**149. Machinery, etc.**—The designing of machinery has not been systematized as has the designing of structural steel work, and the choice of working stresses is to a large extent a matter of individual judgment. The following values are taken from a table in Unwin's Machine Design and are for unvarying stress; if the stress varies, but does not reverse, multiply by two-thirds; if the stress reverses, multiply by one-third.

| | Tension. | Compression. | Torsion. |
|---|---|---|---|
| Cast iron | 4,200 | 12,600 | 2,100 |
| Iron forging | 15,000 | 15,000 | 6,000 |
| Mild steel | 20,000 | 20,000 | 8,000 |
| Cast steel | 24,000 | 24,000 | 10,000 |

For shafting of marine engines 9,000 lb. per sq. in., on wrought iron and 12,000 lb. on steel is commonly used, the shaft being designed for the maximum and not the mean twisting moment. These values are about those used in other machinery.

The working stress in boilers made of 60,000 lb. steel may be as high as 13,000 lb. per sq. in. with first-class workmanship, but ordinarily 10,000 to 11,000 lb. is employed.

# CHAPTER XI.

## INTERNAL STRESS: CHANGE OF FORM.

**150. Introduction.**—Let any body to which external forces are applied be cut by a plane of section. The force with which one part of the body acts against the other part of the body is the *stress* on the plane of section. This stress is distributed over the section and may be uniform or may vary. Its *intensity*, that is, the *unit stress*, at any point is found by dividing the stress on an indefinitely small area surrounding the point by the area. Stresses may be normal, tangential or oblique to the surface on which they act, and they are completely determined only when both intensity and direction are known. It is desirable to know the magnitude and kind of the unit stress at each point in order to be sure that the material can safely resist it; or to determine the required cross-section to reduce the existing stresses to safe values.

A unit stress is expressed as a certain number of pounds of tension, compression, or shear on a square inch of section. If the plane of section is changed in direction, the force on the section may be changed and the area of section may also be changed so that the unit stress on the new section is altered from that on the old in two ways. Stresses per square inch, or unit stresses, therefore cannot be resolved and compounded as can forces, unless they happen to act on the same plane. Generally, each unit stress may be multiplied by the area over which it acts, and the several *forces* so obtained may be compounded or resolved as desired; the final force or forces divided by the areas on which they act will give the desired unit stresses.

Where the stress on a plane varies from point to point, as does the direct stress on the right section of the beam, and as does the shearing stress also in the same case, the investigation is supposed to be confined to so small a portion of the body that the stress over any plane may be considered to be sensibly constant.

**151. Stress on a Section Oblique to a Given Force.**—Suppose a short column or bar to carry a force of direct compression or tension, of magnitude $P$, centrally applied and uniformly distributed over the cross-section, $S$. The unit stress on and perpendicular to the right section will be $p_1 = P \div S$.

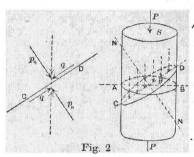

Fig. 2

On an oblique section $CD$, Fig. 2, making an angle $\theta$ with the right section $AB$, the unit stress will be $P \div S \sec \theta = p_1 \cos \theta$, making an angle of $\theta$ with the normal to the oblique section on which it acts. If this oblique unit stress is resolved normally and tangentially to $CD$, the

Normal unit stress $= p_n = p_1 \cos \theta \cdot \cos \theta = p_1 \cos^2 \theta$;
Tangential do. $= q = p_1 \cos \theta \sin \theta$.

The normal unit stress on the oblique plane is of the same kind as $P$, tension or compression; the tangential unit stress, or shear, tends to make one part of the prism slide down and the other part slide up the plane.

The largest normal unit stress for different planes is found when $\theta = 0$, which defines the fracturing plane for tension; the minimum normal unit stress occurs for $\theta = 90°$; and the greatest unit shear is found for $\theta = 45°$, when we have $q_{max.} = \frac{1}{2} p_1$.

**152. Combined Stresses.**—The action line of $P$ may be taken for the axis of X. Two equal and opposite forces, pull or thrust, may then be applied along the axis of Y, and the normal and tangential unit stresses found on the plane just discussed; and similarly for the direction Z. The normal unit stresses, since they act on the same area, may then be added algebraically, and the shearing stresses may be combined; finally a resultant oblique unit stress may be found on the given plane.

Thus, if $p_1$ is the unit stress on a section normal to the X axis, and $p_2$ the unit stress on a section normal to the Y axis,

# INTERNAL STRESS: CHANGE OF FORM. 169

$$p_n = p_1 \cos^2 \theta + p_2 \cos^2(90° - \theta) = p_1 \cos^2 \theta + p_2 \sin^2 \theta,$$
$$q = p_1 \sin \theta \cos \theta - p_2 \cos \theta \sin \theta = (p_1 - p_2) \sin \theta \cos \theta.$$

$p_n$ is a maximum when $\theta = 0$ or $90°$, and $q$ is a maximum when $\theta = 45°$.

$$p_{n\,max.} = p_1 \text{ or } p_2;$$
$$q_{max.} = \tfrac{1}{2}(p_1 - p_2).$$

A more convenient method will, however, be developed and used in the following sections. As most of the forces which act on engineering structures lie in one plane or parallel planes, such cases chiefly will be considered.

**153. Unit Shears on Planes at Right Angles.**—If, in the preceding illustration, the unit stresses, both normal and tangential, are found on another plane N which makes an angle $\theta' = 90° - \theta$ with the right section, there will result

$$p_n' = p_1 \cos^2 \theta' = p_1 \sin^2 \theta; \qquad q' = q.$$

Hence, on a pair of planes of section, both of which are at right angles to the plane of external forces and to each other, the *unit shears are of equal magnitude*, and, since $p_n + p_n' = p_1$, the unit normal stresses are together equal to the original normal unit stress. It is further evident that one normal unit stress $p_n'$ may be found by subtraction as soon as the other is known, and that ordinary resolution on these two planes of the original unit stress would be erroneous.

**154. Unit Shears on Planes at Right Angles Always Equal.**—Since, as before stated, other forces, in other directions, may be simultaneously applied to the given body, and their effects found on the same two planes, it follows that, in any body under stress, equal unit shearing stresses will exist on two planes each of which is perpendicular to the direction of the shear on the other.

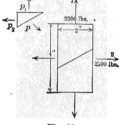

Fig. 61

*Example.*—A closed cylindrical receiver, ¼ in. thick, has a spiral riveted joint making an angle of 30° with the axis of the cylinder, and a portion 2 in.×4 in. of the cylinder, Fig. 61, has the given tensions of 2,500 lb. acting upon it. Then $p_1 = 2,500 \div 2 \cdot \frac{1}{4} = 5,000$ lb. per sq. in., and $p_2 = 2,500 \div 4 \cdot \frac{1}{4} = 2,500$ lb. per sq. in.

$$p_n = 5,000 \cdot \tfrac{3}{4} + 2,500 \cdot \tfrac{1}{4} = 4,375 \text{ lb. per sq. in.,}$$
$$q = 5,000 \cdot 0.433 - 2,500 \cdot 0.433 = 1,082 \text{ lb. per sq. in.,}$$
$$p = \sqrt{(4,375^2 + 1,082^2)} = 4,507 \text{ lb. per sq. in.,}$$

or $4,507 \cdot \frac{1}{4} = 1,127$ lb. per linear inch of joint, which value will determine the necessary pitch of the rivets for strength.

The stress on a joint at right angles to the above can be similarly found. An easier process will be given in § 161.

**155. Compound Stress** is the internal state of stress in a body caused by the combined action of two or more simple stresses (or balanced sets of external forces) in different directions, as in the above example. The investigations which follow are those of compound stress, but they will, as above stated, be chiefly confined to stresses in or parallel to one plane.

**156. Conjugate Stresses: Principal Stresses.**—If the stress on a given plane in a body is in a given direction, the stress on a

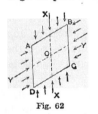

Fig. 62

plane *parallel* to that direction must be parallel to the first-mentioned plane. For the equal resultant forces exerted by the other parts of the body on the faces A B and C D of the prismatic particle, Fig. 62, are *directly opposed* to one another, their common line of action traversing the axis of X through O; and they are therefore independently balanced. Therefore the forces exerted by the other parts of the body on the faces A D and B C of this prism must be independently balanced and have their resultants directly opposed; which cannot be unless their direction is parallel to the plane Y O Y.

A pair of stresses, each acting on a plane parallel to the direction of the other, is said to be *conjugate*. Their unit values are independent of each other, and they may be of the same or opposite kinds. If they are normal to their planes, and hence at right angles to each other, they are called *principal* stresses.

*Examples.*—The unit stress found in § 154 makes an angle with the plane on which it acts whose tangent is $4,375 \div 1,082 = 4.04$. Upon a new plane cutting the metal in this direction the stress must act in a direction parallel to the joint referred to.

If a plane be conceived parallel to a side-hill surface, at a given vertical distance below the same, the pressure at all points of that plane, being due to the weight of the prism of earth above any square foot of the plane, will be vertical and uniform. Then must the pressure on a vertical plane transverse to the slope be parallel to the surface of the ground. That the pressure against the vertical plane is not horizontal, but inclined in the direction stated, is shown by the movement of sewer trench sheeting and braces, when the braces are not inclined up hill, but are put in horizontally.

**157. Shearing Stress.**—If the stresses on a pair of planes are *entirely* tangential to those planes, the unit shears must be equal. Consider them as acting along the faces of a small prismatic particle A B C D, which lies at O. The moment of the total shear on the two faces A B and C D must balance the moment for the faces A D and B C, for equilibrium.

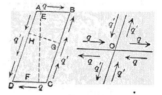

Fig. 63

$$q \cdot A B \cdot E F = q' \cdot A D \cdot H G.$$

But the area of A B C D, A B·E F = A D·H G; ∴ $q' = q$.

This construction shows further that a shear cannot act alone as a simple stress, but must be combined with an equal unit shear on a different plane.

**158. Two Equal and Like Principal Stresses.**—If a pair of *principal* stresses, § 156, are equal unit stresses of the same kind, $p_1$ and $p_2$, Fig. 64, the stress on *every* plane is *normal* to that plane, and of the same kind and magnitude.

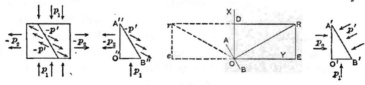

Fig. 64

Let $p_1$ act in the direction O X on the plane O' B' of the prismatic particle O' A' B' which lies at O, and $p_2$ act in the direc-

172    STRUCTURAL MECHANICS.

tion O Y on the plane O' A', $p_1$ and $p_2$ being equal unit stresses of the same kind. Make $O D = p_1 \cdot O' B'$, the total force on O'B', and $O E = p_2 \cdot O' A'$, the total force on O' A', both being positive. Complete the rectangle O D R E. Then must R O, applied to the plane A'B', be necessary to insure equilibrium of the prism O' A' B'. Hence $p' = R O \div A' B'$. Since $p_1 = p_2$,

$$\frac{OD}{O'B'} = \frac{OE}{O'A'} = \frac{OR}{A'B'}; \quad \therefore p' = p_1 = p_2.$$

Because of similarity of triangles A' O' B' and O E R, R O is perpendicular to A B, or $p'$ to A' B', and is of the same kind as $p_1$ and $p_2$.

*Example.*—Fluid pressure is normal to every plane passing through a given point, and equal to the pressure per square inch on the horizontal plane traversing the point. Here manifestly the three coordinate axes of X, Y, and Z might be taken in any position, as all stresses are principal stresses.

**159. Two Equal and Unlike Principal Stresses.**—If a pair of *principal* stresses are equal unit stresses of opposite kinds, as $p_1$ and $-p_2$, Fig. 64, the unit stress on every plane will be the same in magnitude, but the angle which its direction makes with the normal to its plane will be bisected by the axis of principal stress, and its kind will agree with that of the principal stress to which it lies the nearer.

In this case lay off Oe in the negative direction, to represent $-p_2 \cdot O'' A''$; construct the rectangle O Dre, and draw rO which will be the required force distributed over A'' B'' to balance the forces O'' B'' and O'' A''. This force rO will be the same in magnitude as R O, making $p' = p_1 = p_2$ and rO will make the same angle with O X or O Y as R O does. As R O lies on the normal to A B, and O X bisects R Or, the statement as to position is proved. The stress $p'$ agrees in kind with that one of the principal stresses to which its direction is nearer.

**160. Two Shears at Right Angles Equivalent to an Equal Pull and Thrust.**—If the plane A'' B'' is at an angle of 45° with O X, rO will coincide with A B and becomes a shear. Therefore two

# INTERNAL STRESS: CHANGE OF FORM. 173

equal unit stresses of opposite kinds, that is a pull and a thrust and normal to planes at right angles to one another, are equivalent to, and give rise to equal unit shears on planes making 45° with the first planes and hence at right angles to each other; and *vice versa*.

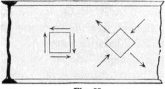

Fig. 65

*Example.*—If, at a point in the web of a plate girder, Fig. 65, there is a unit shear, and nothing but shear, on a vertical plane, of .4,000 lb., there must be a unit shear of 4,000 lb., and nothing but shear, on the horizontal plane at that point and on the two planes inclined at 45° to the vertical through the same point there will be, on one, only a normal unit tension of 4,000 lb., and on the other an equal normal unit thrust.

**161. Stress on any Plane, when the Principal Stresses are Given.**—This general problem is solved by dividing it into two special cases; the one that of two equal and like principal stresses, the other that of two equal and unlike principal stresses. Let the two principal unit stresses be $p_1 = $ O D, and $p_2 = $ O F, of any magnitude, and of the same kind or opposite kinds. Fig. 66. The direction, magnitude and kind of the unit stress on any plane A B is desired.

Let $p_1$ be the greater. Divide $p_1$ and $p_2$ into their half sum and difference as follows:

$$p_1 = \tfrac{1}{2}(p_1 + p_2) + \tfrac{1}{2}(p_1 - p_2), \quad \text{and} \quad p_2 = \tfrac{1}{2}(p_1 + p_2) - \tfrac{1}{2}(p_1 - p_2).$$

The distance O C or O E will represent the half sum $\tfrac{1}{2}(p_1 + p_2)$, and C D or E F the half difference $\tfrac{1}{2}(p_1 - p_2)$. If $p_1$ and $p_2$ are the same sign the right-hand figure will result; if of opposite signs, the left-hand figure will be obtained.

By the principle of § 158, when the two equal principal unit stresses O C and O E are considered, lay off O M on the normal to the plane whose trace is A B, for the direction and magnitude of the unit stress on A B due to $\tfrac{1}{2}(p_1 + p_2)$. There remain C D and E F representing $+\tfrac{1}{2}(p_1 - p_2)$ on the vertical axis, and $-\tfrac{1}{2}(p_1 - p_2)$ on the horizontal axis respectively.

174                    STRUCTURAL MECHANICS.

By § 159, lay off O Q, making the same angle with O X as does O M, but on the opposite side, or in the contrary direction, for the magnitude and direction of stress on plane A B due to $\pm \frac{1}{2}(p_1 - p_2)$. As O M and O Q both act on the same unit of area of A B, R O, in the opposite direction to their resultant O R, will give the direction and magnitude of the unit stress on A' B' to balance $p_1$ on O' B' and $p_2$ on O' A'. In the figure on the right R O is positive, or compression. If, in the figure on the left, where $p_1$ is + and $p_2$ is —, R O falls so far to the left as to come on the other side of A B, it will agree with $p_2$, and

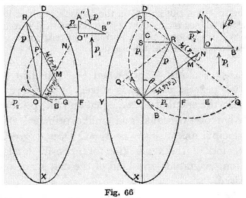

Fig. 66

be negative or tension. If A B is taken much more steeply inclined, such will be the case. The small prisms illustrate the constructions. If R O falls on A B, it will be shear. Some constructions for different inclinations of plane A B will be helpful to an understanding of the matter.

A much abbreviated construction is as follows:—Strike a semicircle from M on the normal, with a radius M O = $\frac{1}{2}(p_1 + p_2)$, and draw M R through the points where the semicircle cuts the axes of $p_1$ and $p_2$. The angle N M R is thus double the angle M O D, since it is an exterior angle at the vertex of an isosceles triangle. Lay off M R = $\frac{1}{2}(p_1 - p_2)$ in the direction of the axis of greatest stress, and R O will be the desired unit stress on A B. If $p_2$ is opposite in kind to $p_1$, M R will be greater than M O, and R will go beyond P.

## INTERNAL STRESS: CHANGE OF FORM.

**162. Ellipse of Stress.**—For different planes A B through O, $p_1$ and $p_2$ being given, the locus of M is a circle of radius $\frac{1}{2}(p_1+p_2)$, and the locus of R is an ellipse (as will be proved below), with major and minor semidiameters $p_1$ and $p_2$. Hence it is seen why $p_1$ and $p_2$, normal to the respective planes, and at right angles, are called *principal* stresses.

If three principal stresses, coinciding in direction with the rectangular axes of X, Y and Z, simultaneously act on a given point, an ellipsoid constructed on them as semidiameters will limit and determine all possible stresses on the various planes which can be passed through that point in the body.

That the locus of R is an ellipse may be proved as follows: Drop a perpendicular R S from R on O X. P M O and O M G are isosceles triangles. $<$ P O M $=<$ M P O $=\theta$.

$$O M = M P = \tfrac{1}{2}(p_1+p_2); \quad G R = p_1; \quad P R = p_2.$$
$$R S = P R \sin M P O = p_2 \sin \theta,$$
$$S O = G R \cos M P O = p_1 \cos \theta,$$
$$O R = p_r = \sqrt{(S O^2 + R S^2)} = \sqrt{(p_1^2 \cos^2 \theta + p_2^2 \sin^2 \theta)}, \quad . \quad (1)$$

which is the value of the radius vector of an ellipse in terms of the eccentric angle, the origin being at the centre.

As $<$ N M R $= 2 \cdot <$ P O M $= 2\theta$,
$\sin$ M O R : $\sin$ O M R $=$ R M : O R $= \tfrac{1}{2}(p_1-p_2) : p_r$;

$$\therefore \sin M O R = \sin 2\theta \cdot \frac{p_1-p_2}{2 p_r}, \quad \ldots \quad (2)$$

which gives the obliquity of the unit stress to the normal to the plane, in terms of the angle of the normal with the axis of greater principal stress, or of the plane with the other axis. The graphical construction gives the stress and its angle with the normal or the plane by direct measurement, and is far more convenient and less liable to error.

If $p_1 = p_2$, the case reduces to that of § 158 or § 159.

If the ellipse whose principal semidiameters are $p_1$ and $p_2$ is given, the unit stress on any plane may briefly be found by drawing the *normal* to the plane, laying off O M $= \tfrac{1}{2}(p_1+p_2)$, taking a radius of $\frac{1}{2}(p_1-p_2)$, and, with M as a centre, cutting the ellipse at R on the side of the normal towards the greater axis. A line R O will be the desired unit stress.

*Example.*—Let $p_1=100$ lb. on sq. in., $p_2=-50$ lb. Plane A B makes 30° with direction of $p_2$, or its normal makes 30° with $p_1$. Construct the figure and find the magnitude, direction, and sign of the unit stress on the oblique plane. Try other values.

**163. Shearing Planes.**—To determine the angle of the planes on which there is only shear, and the conditions which render shearing planes possible.

If the plane A B of the previous figure is to be the shearing plane, there must be no normal component upon it, and therefore from § 151, if the plane makes an angle $\theta_s$ with $p_2$, or the normal to it is inclined at an angle $\theta_s$ to $p_1$,

$$p_n = p_1 \cos^2 \theta_s + p_2 \sin^2 \theta_s = 0.$$

$$\therefore \frac{\sin \theta_s}{\cos \theta_s} = \tan \theta_s = \sqrt{\frac{-p_1}{p_2}}.$$

No shearing plane is possible unless $p_1$ and $p_2$ differ in sign. There will then be two planes of shear making equal angles with the direction of $p_2$ or of $p_1$.

In the above example, $\sqrt{(100 \div 50)} = \sqrt{2} = \tan \theta_s = 1.4142$. $\theta = 54° 44'$.

If the ellipse of stress is drawn, take a radius equal to the side of a right angled triangle whose other side is $\frac{1}{2}(p_1+p_2)$, and hypothenuse is $\frac{1}{2}(p_1-p_2)$, and strike a circle from the centre of the ellipse. Planes drawn through the points of intersection of this circle and ellipse and the centre will be the shearing planes. Unless $p_1$ and $p_2$ differ in sign, the circle will be imaginary. The value of the shear on these planes is

$$q = \sqrt{(\tfrac{1}{4}(p_1-p_2)^2 - \tfrac{1}{4}(p_1+p_2)^2)} = \sqrt{-p_1 p_2}.$$

**164. Given any Two Stresses: to Find Principal Stresses.**—As, in actual practice, two oblique unit stresses on different planes may often be known in magnitude, direction and sign, it will be required to find the principal unit stresses, since one of them is

## INTERNAL STRESS: CHANGE OF FORM.     177

the maximum stress to be found on any plane, and the other is the minimum stress of the same kind, or the maximum normal stress of the opposite kind.

Given two existing unit stresses, $p$ and $p'$ of any direction, magnitude and sign, to determine the principal unit stresses, $p_1$ and $p_2$.

If $p_1$ and $p_2$ were known, and $p$ and $p'$ were then to be found from the former, the construction shown in Fig. 67 would be made, in which $O\,M = O\,M' = \tfrac{1}{2}(p_1 + p_2)$ and $M\,R = M'\,R' = \tfrac{1}{2}(p_1 - p_2)$. If one of these normals were revolved around O to coincide with the other, the point M' would fall on M, but M' R' would diverge from M R, while equal in length to it.

Hence, when $p$ and $p'$ are the given quantities, let A B, Fig. 68, represent the trace of the plane on which $p$ acts, O N the

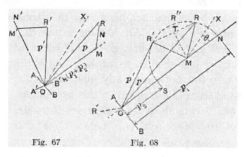

Fig. 67     Fig. 68

normal to that plane, and O R the unit stress $p$ in magnitude and actual direction of action on A B. O R represents either tension or compression, as the case may be. Now let the plane on which $p'$ acts, together with its normal and $p'$ itself in its relative position, be revolved about O until it coincides with A B. Its normal will fall on O N and $p'$ will be found at O R', on one side or the other of O N, if it is of the same kind with $p$; or it is to be laid off on the dotted line below, if of the opposite sign.

In other words, lay off $p'$ from O, at the same angle with O N which it makes with the normal to its own plane. It is well, for accuracy of construction, to draw it on the same side of the normal as $p$, the result being the same as if it were drawn on the other side. (The change from one side of the normal to the other simply consists in using the corresponding line on the other side of the main axis of the ellipse of stress). Thus is found O R' or −O R' as the case may be. Draw R R' and, since R M R' must be an isosceles triangle, bisect R R' at T and drop a perpendicular to R R' from T on to the normal, cutting the latter at M. Then since, as previously pointed out, O M = $\frac{1}{2}(p_1+p_2)$ and M R = M R' = $\frac{1}{2}(p_1-p_2)$,—with M as a centre, and radius M R, describe a semicircle; O N will be $p_1$ and O S will be $p_2$. Since $p$ is in its true position, and the angle N M R = 2 M O D of Fig. 66 or 2 M O X of Fig. 67, the direction of the axis X along which $p_1$ acts will bisect N M R, and the axis along which $p_2$ acts will be perpendicular to axis X. They may be now at once drawn through O, if desired.

**165. From any Two Stresses to Find Other Stresses.**—From the preceding construction, § 164, the stress on any other plane may now be found. All *possible* values of $p$ consistent with the two, O R and O R', first given, will terminate, in Fig. 68, on the semicircle just drawn, as at R'', and the greatest possible obliquity to the normal to any plane through O will be found by drawing from O a tangent to this semicircle.

**166. When Shearing Planes are Possible.**—In case the lower end of the semicircle cuts below O, Fig. 69, $p_1$ and $p_2$ are of opposite signs, all obliquities of stress are possible, and the distance from O to the point where the semicircle cuts A B, being

Fig. 69

perpendicular to the normal O N, gives the unit shear on the shearing planes. If $p_1$ and $p_2$ are drawn through O in position, and the ellipse of stress is then constructed on them as semi-diameters (as can be readily done by drawing two concentric circles with $p_1$ and $p_2$ respectively as radii, and projecting at right angles, parallel to $p_1$ and $p_2$, to an intersection, the two points where any radius cuts both circles), an arc described from O, with a radius equal to this unit shear and cutting the ellipse, will locate a point in the shearing plane which may then be drawn through that point and O. Two shearing planes are thus given, as was proved to be necessary, § 157.

The above solution may be considered the general case.

# INTERNAL STRESS: CHANGE OF FORM.

**167. From Conjugate Stresses to Find Principal Stresses.—** If $p$ and $p'$ are conjugate stresses, it is evident, from definition, and from Fig. 62, that they are equally inclined to their respective normals. Hence O R' will fall on O R, when revolved, both O R and O R' lying above O when of the same sign, and on opposite sides of A B when of opposite signs. The rest of the construction follows as before, being somewhat simplified.

It may be noted that, when $p_1 = \pm p_2$, the propositions of §§ 158 and 159 are again illustrated.

One who is interested in a mathematical discussion of this subject is referred to Rankine's "Applied Mechanics," where it is treated at considerable length. This graphical discussion is much simpler, less liable to error, and determines the stresses in their true places.

**168. Stresses in a Beam** —The varying tensile and compressive stresses on any section of a beam are accompanied by varying shearing stresses on that section and by equal shears on a longitudinal section. The direct or normal stresses due to bending moment vary uniformly from the neutral axis either way, § 62; the shears are most intense at the neutral axis, § 72. The normal and shearing stresses on the cross-section of a rectangular beam and also the resultant stresses on the section are shown in

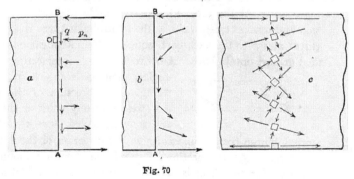

Fig. 70

Fig. 70, $a$ and $b$. The maximum and minimum unit stresses (that is, the principal stresses) at any point, with their directions, may be found by a slight modification of § 164 as follows:

Let Fig. 71 represent the stresses acting on a particle, as the

one at O, Fig. 70. Lay off $ON = p_n$ on the normal to the plane A B, on which it acts in its true position. As $q$ acts on the same

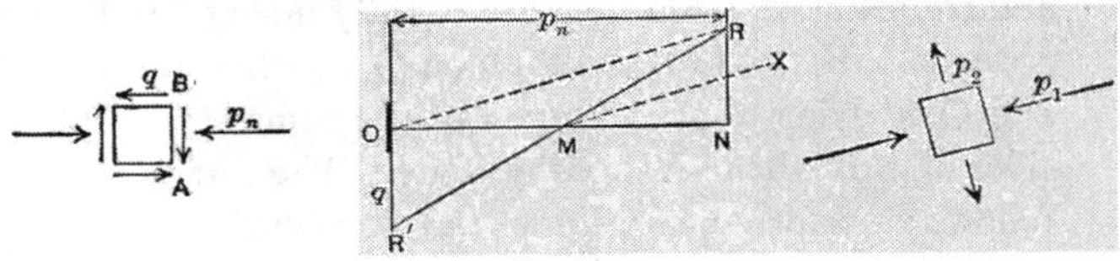

Fig. 71

area of A B as does $p_n$, lay off $RN = q$ at right angles. The two component stresses being R N and N O, a line from R to O will represent the direction and magnitude of the unit stress on A B. Revolve the horizontal plane through 90° to coincide with A B. Then lay off $R'O = q$ as shown. (As explained in § 164, R' O may be laid off on the other side of the normal, O N, but that construction is not quite so simple.) Since R O and R' O represent $p$ and $p'$ of the preceding sections, connect R and R' and bisect R R' at M, which point falls on O N and is also the point where the perpendicular from R R' will strike O N. Hence $p_1 = OM + MR$ and $p_2 = OM - MR$ and the direction of $p_1$ is M X, which bisects $< RMN$. The principal stresses are shown in the figure.

If the principal stresses are found at various points on the cross-section of a loaded beam, the results will be as shown in Fig. 70, c. At the top and bottom simple stresses of compression and tension exist, while at the neutral axis there is a pull and thrust at 45° to the axis, each equal to the unit shear on the vertical or horizontal plane. A network of lines intersecting at right

Fig. 72

angles, such as is shown in Fig. 72, will give the direction of the principal stresses at any point of the beam and may be called *lines of principal stress*. Those convex upward give the direction of the compressive forces. At the section of maximum bending moment, at which section the shear is zero, the curves are horizontal and the stress is greater than at any other point on the curve; from there the stress diminishes to zero at the edge of the beam. The lines of principal stress also show the traces of surfaces on which the stress is always normal.

## INTERNAL STRESS: CHANGE OF FORM.

**169. Rankine's Theory of Earth Pressure.** — Any embankment or cutting keeps its figure both by the friction and by the adhesion between the grains of earth. The latter is variable and uncertain in amount and may even be entirely absent, for which reason it is neglected in the present discussion. A bank of earth is, therefore, to be conceived as a granular mass, retaining its shape by friction alone, as is the case with a pile of dry sand. When any such granular material is heaped up in a pile, the sides slide until they assume a certain definite slope depending upon the material. The greatest angle with the horizontal which the sides can be made to take is the *angle of repose*, and its tangent is the coefficient of friction of the material.

It is evident that in a granular mass there is pressure on any plane which can be passed through it, and from § 161 it is seen that, if the principal stresses in the mass are unequal, the stress on any plane, not a principal plane, is oblique. But in earth, the obliquity of the pressure on any plane cannot exceed the angle of repose, $\phi$, or sliding would take place along that plane. If

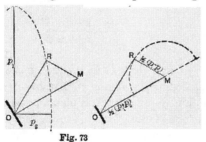

Fig. 73

Fig. 73 shows the plane upon which the direction of pressure is most oblique (see § 165), it is necessary for equilibrium that the angle R O M shall not exceed $\phi$. Then

$$\sin R\,O\,M = \frac{p_1 - p_2}{p_1 + p_2} \leq \sin \phi.$$

By rearranging the terms, the ratio between the principal pressures is found to be

$$\frac{p_2}{p_1} \geq \frac{1 - \sin \phi}{1 + \sin \phi}.$$

In the figure, $p_1$ is taken larger than $p_2$; if $p_2$ is larger than $p_1$, the triangle will lie on the other side of the normal and the ratio becomes

$$\frac{p_2}{p_1} \leq \frac{1+\sin\phi}{1-\sin\phi}.$$

If $p_1$ is the principal pressure acting vertically and due to the weight of the earth itself, its conjugate pressure, $p_2$, must lie between the values given by the two preceding equations. Its exact amount can be determined by the principle of least resistance, for $p_1$, which is caused by the action of gravity, produces a tendency for the earth to spread laterally, which tendency gives rise to the pressure, $p_2$. As $p_2$ is caused by $p_1$, $p_2$ will not increase beyond the least amount sufficient to balance $p_1$, hence

$$\frac{p_2}{p_1} = \frac{1-\sin\phi}{1+\sin\phi},$$

and in Fig. 73 the angle R O M is equal to $\phi$.

In the preceding discussion $p_1$ has been considered to act on a horizontal, and $p_2$ on a vertical plane, and such will be the case when the surface of the ground is level, but if the ground slopes, the pressure on a plane parallel to the surface is vertical and the direction of $p_1$ inclines slightly from the vertical, as shown in the following example:

*Example.*—Find the pressure on the back of the retaining-wall shown in Fig. 74, and also the resultant pressure on the joint C O'. The pressure on the plane O O' passed parallel to the surface of the ground is vertical and due to the weight of the earth upon it of depth K O. But the prism of earth resting on one square foot of the plane has a smaller horizontal section than one square foot, and the ratio of the unit vertical pressure on the plane through O to the weight of a vertical column of earth one square foot in cross-section will be that of the normal O H to the vertical O K. Hence O R(= O H) represents in feet of earth the pressure per square foot on the plane O O'. Draw a line, O S, making the angle of repose, $\phi$, with the normal and by trial find on O H a centre, M, from which may be drawn a semicircle tangent to the line, O S, and passing through R. Then O M = $\frac{1}{2}(p_1 + p_2)$ and

### INTERNAL STRESS: CHANGE OF FORM. 183

$M R = \frac{1}{2}(p_1 - p_2)$, and M X gives the direction of $p_1$, the unit of pressure being the weight of one cubic foot of earth.

The principal stresses are now known and the pressure on the back of the wall can be found. To find the pressure on O' B at O', draw O' A parallel to M X and O' M' normal to the back of the wall. Lay off O' M' equal to O M, and M' R' equal to M R and making the same angle with O' A as O' M' does. Then R' O' gives the direction of pressure on the back of the wall and, when measured by the scale of the

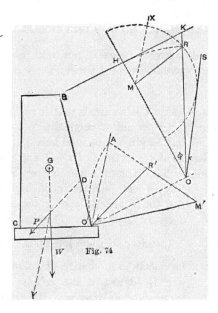

Fig. 74

drawing and multiplied by the weight of a cubic foot of earth, gives the pressure in pounds per square foot at O'.

As the pressure on the back increases regularly with the distance below the surface of the ground, the centre of pressure will be at D, one-third of the slant height from O', and the total earth pressure against one foot in length of the wall will be $P = \frac{1}{2} \times O' B \times O' R' \times$ weight of a cubic foot of earth. If $W$ is the weight of one foot length of wall applied at the centre of gravity, G, the resultant of $P$ and $W$ is the resultant pressure on the bed-joint, O' C, and the point where it cuts the joint should lie within the middle third.

**170. Change of Volume.**—If $p =$ unit stress per square inch on the cross-section of a prism, and $\lambda$ is the resulting stretch or shortening *per unit of length*, then by definition $E = p \div \lambda$, if $p$ does not exceed the elastic limit.

When a prism is extended or compressed by a simple longitudinal stress, it contracts or expands laterally, Fig. 75. This contraction or expansion per unit of breadth may be written $\mp \lambda \div m$ in which $m$, the ratio of longitudinal extension to lateral contraction, is a constant for a given material, and for most solids lies between 2 and 4.

A simple longitudinal tension $p$ then accompanies a

Longitudinal stretch $= \lambda = p \div E$ per unit of length, and a
Transverse contraction $= -\lambda \div m = -p \div mE$ per unit of breadth.

For ordinary solids $\lambda$ is so small that it makes no difference whether it is measured per unit of original or per unit of stretched length. The original length will be used here.

The new length of the prism is $l(1+\lambda)$ and the cross-section is $S(1-\lambda \div m)^2$. The volume has changed from $Sl$ to $Sl(1+\lambda-2\lambda \div m)$ nearly, if higher powers of $\lambda$ than the first are dropped, since the unit deformations are very small. The change of unit volume is therefore $\lambda\left(1-\dfrac{2}{m}\right)$. Thus, if $m$ is nearly 4, for metals, the change of volume of one cubic unit is $\frac{1}{2}\lambda$ nearly, the volume being *increased* for longitudinal tension. If there were no change of volume, $m$ would be 2, as is the case for india rubber for small deformations. Similarly, for compression the change of the unit volume is nearly $-\frac{1}{2}\lambda$ for metals, the volume being *diminished*.

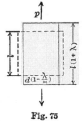
Fig. 75

*Example.*—Steel, $E = 29{,}000{,}000$; $p = 20{,}000$ lb. per sq. in. tension; the extension will be $\dfrac{1}{1{,}450}$ of its initial length, the lateral contraction will be about $\dfrac{1}{5{,}800}$ of its initial width, and its increase of volume about $\dfrac{1}{2{,}900}$.

## 171. Effect of Two Principal Stresses.

Denote the stresses by $p_1$ and $p_2$, treated as tensile. If they are compressive, reverse the signs.

Under the action of $p_1$ there will be the following stretch of the sides per unit of length: Fig. 76,

parallel to O C, $\quad \dfrac{p_1}{E}$;

parallel to OB and O A, $-\dfrac{p_1}{mE}$.

Under the action of $p_2$ there will be

parallel to O B, $\quad \dfrac{p_2}{E}$;

parallel to O A and O C, $-\dfrac{p_2}{mE}$.

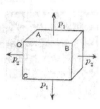

Fig. 76

Adding the parallel changes or stretches

parallel to O C, $\lambda_1 = \dfrac{1}{E}\left(p_1 - \dfrac{p_2}{m}\right)$;

parallel to O B, $\lambda_2 = \dfrac{1}{E}\left(p_2 - \dfrac{p_1}{m}\right)$;

parallel to O A, $\lambda_3 = -\dfrac{1}{mE}(p_1 + p_2)$.

If $p_1$ and $p_2$ are equal unit stresses, but of opposite signs, the changes of length become

$$\dfrac{p}{E}\left(1 + \dfrac{1}{m}\right); \quad -\dfrac{p}{E}\left(1 + \dfrac{1}{m}\right); \quad \text{and zero};$$

or, putting either of these two changes equal to $\lambda$, the lengths of the sides of the cube originally unity per edge will be $1 + \lambda$, $1 - \lambda$, and $1$, and the volume, neglecting $\lambda^2$, is unchanged.

**172. Effect of Two Shears.**—The cube shown in Fig. 77 has been deformed by the action of two equal and opposite principal stresses, and a square, traced on one side of the original cube, has been distorted to a rhombus, the angles of which are greater and less than a right angle by the amount, $\phi$. By § 160, two equal principal stresses of opposite sign are equivalent to two unit shears of the same amount per square inch on planes at 45° with the principal axes; hence the distortion of the prism, whose face is the rhombus, results from two equal shears at right angles.

Now one-half the angle $\frac{1}{2}\pi - \phi$ has for its tangent $\frac{1}{2}(1-\lambda) \div \frac{1}{2}(1+\lambda)$; hence

$$\frac{1-\lambda}{1+\lambda} = \tan \tfrac{1}{2}(\tfrac{1}{2}\pi - \phi) = \frac{1-\tan \tfrac{1}{2}\phi}{1+\tan \tfrac{1}{2}\phi}; \quad \text{or} \quad \lambda = \tan \tfrac{1}{2}\phi.$$

But as $\phi$ is small, $\lambda = \tfrac{1}{2}\phi$, or $\phi = 2\lambda$.

Therefore a stretch and an equal shortening, along a pair of rectangular axes, are equivalent to a simple distortion relatively to a pair of axes making angles of 45° with the original axes; and the amount of distortion is double that of either of the direct changes of length which compose it. This fact also appears from the consideration that a distortion of a square is equivalent to an elongation of one diagonal and a shortening of the other in equal proportions.

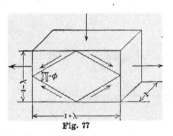

Fig. 77

*Example.*—For steel, as before, $\lambda = \dfrac{1}{1,450}$, $\phi = \dfrac{1}{725} = 4'\ 45''$, if $p_1 = -p_2 = 20,000$ lb.

**173. Modulus of Shearing Elasticity.**—Similarly, equal shearing stresses $q$ on two pairs of faces of a cube, in directions parallel to the third face, will distort that third face into a rhombus, each angle being altered an amount $\phi$, there being distortion of shape only, and not change of volume, Fig. 78.

Under the law which has been proved true within the elastic limit, and the definition of the modulus of elasticity, § 10, a modulus of transverse (or shearing) elasticity, $C$, also called coefficient of rigidity, as $E$ may be called coefficient of stiffness, may be written, $C = q \div \phi$.

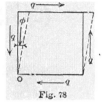

Fig. 78

As these two unit shears are equivalent to a unit pull and thrust of the same magnitude per square inch, at right angles with each other and at 45° with these shears, the case is identical with the preceding one. Then

$$\phi = 2\lambda, \quad \text{and} \quad \lambda = \frac{p}{E}\left(1 + \frac{1}{m}\right). \quad \therefore \quad \phi = \frac{2p}{E} \cdot \frac{m+1}{m}.$$

But, as $p = q = C\phi$, $\quad C = \dfrac{p}{\phi} = \dfrac{1}{2} \cdot \dfrac{mE}{m+1}$.

For iron and steel $m$ is nearly 4, which gives $C = \frac{2}{5}E$. For wrought iron and steel, $C$ is one or two one-hundredths less than $0.4E$. Some use $\frac{3}{8}E$. $C = 11{,}000{,}000$ is a fair value.

**174. Stress on One Plane the Cause of Other Stresses.**—The elongation produced by a pull, the shortening produced by a thrust, and the distortion due to a shear can be laid off as graphical quantities and discussed as were unit stresses themselves. All the deductions as to stresses have their counterparts in regard to changes of form. There has been found an ellipse of stress for forces in one plane, when two stresses are given. Also, when three stresses not in one plane are given, there is an ellipsoid of stress which includes all possible unit stresses that can act on planes in different directions through any point in a body. So there is an ellipse or ellipsoid that governs change of form.

Whether the movement of one particle towards, from or by its neighbor sets up a resisting thrust, pull or shear, or the application of a pressure, tension, or shear is considered to cause a corresponding compression, extension, or distortion, the stresses and the elastic change of form coexist. Hence it follows that, when a bar is extended under a pull and is diminished in lateral dimensions, a compressive stress acting at right angles to the pull must be aroused between the particles, and measured per unit of area of longitudinal planes, together with shears on some inclined sections.

That such a state of things can exist may be seen from the following suggestions. It may be conceived that the particles of a body are not in absolute contact, but are in a state of equilibrium from mutual actions on one another. They resist with increasing stress all attempts to make them approach or recede from each other, and, if the elastic limit has not been exceeded, they return to their normal positions when the external forces cease to act. The particles in a body under no stress may then be conceived to be equidistant from each other. The smallest applied external force will probably cause change in their positions.

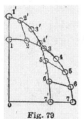
Fig. 79

If, in the bar to which tension is to be applied, a circle is drawn about any point, experiment and what has been stated about change of form in different directions show, that the diameter in the direction of the pull will be lengthened when the force is applied, the diameter at right angles will be shortened, and the circle will become an ellipse. In Fig. 79, particle 1 moves to 1', 2 to 2', 4 to 4', and 7 to 7'. As they were all equidistant from o in the beginning, 1 in moving to 1' offers a tensile resistance, 7 resists the tendency to approach o, while a particle near 4, moving to 4', does not change its distance from o, but moves laterally setting up a shearing stress. A sphere will similarly become an ellipsoid.

**175. Equivalent Simple Stress.**—Our knowledge of the strength of materials is derived largely from experiments upon test pieces in which the load is applied in a direction along the piece—that is, from pieces in a condition of simple stress—and, likewise, working stresses are generally considered as simple stresses. The question then arises whether a body which is stressed in two directions—that is, a body in a condition of compound stress from the action of two principal stresses—is stronger or weaker than if stressed in one direction only.

For example, consider the stresses acting in the shell of a boiler of radius $r$ and steam pressure $p$. §§ 192 and 194 show that on any longitudinal section there is a tension of $pr$ per unit of length, and on a circumferential section there is a tension of $\frac{1}{2}pr$ per unit of length. The tension $pr$ produces an elongation in the direction of the circumference of $\frac{pr}{Et}$ per unit of length, if $t$ is the thickness of the plate. But by § 170 the tension of $\frac{1}{2}pr$ acting at right angles to $pr$ causes the plate to shorten in a circumferential direction an amount $\frac{1}{4} \cdot \frac{\frac{1}{2}pr}{Et}$ per unit of length, if $m$ is 4 for metals. The resulting lengthening of the plate circumferentially under the influence of the two principal stresses is, therefore, but $\frac{7}{8}\frac{pr}{Et}$ per unit of length. The simple stress on unit length of plate which would produce this stretch is $\frac{7}{8}pr$, and if the deformation of the material is considered to be a measure of its strength, the boiler is stronger because of the stress on a circumferential joint than it would be if such stress did not exist.

Hence, if three principal stresses, $p_1$, $p_2$, $p_3$, exist at a point in a body, by § 171 the *equivalent simple stress* which would produce the same deformation at that point in the direction in which $p_1$ acts is

$$p_1' = p_1 - \frac{1}{m}(p_2 + p_3).$$

If there are but two principal stresses as in most of the problems discussed in this chapter, $p_3$ is zero. If $p_2$ and $p_3$ are of opposite kind to $p_1$, the equivalent stress $p_1'$ is greater than $p_1$.

An application of equivalent stress to design is given in § 86. The stresses in plates of §§ 220 and 221 are equivalent stresses.

*Example.*—At a certain point in a conical steel piston there exist principal stresses of 3,160 and 1,570 lb., of opposite signs. Then $p_1' = 3,160 + \frac{1}{4} \cdot 1,570 = 3,550$ lb.; $p_2' = 1,570 + \frac{1}{4} \cdot 3,160 = 2,360$ lb.

**176. Cooper's Lines.**—Steel plate as it comes from the mill has a firmly adhering but very brittle film of oxide of iron on the surface. This film is dislodged by the extension of a test specimen in tension when the yield-point is passed. If a hole is punched at moderate speed in a steel plate, so that the particles under the punch have some opportunity to flow laterally under the compression, there will be a radial compressive stress in all directions outwardly from the circumference of the hole, and a tensile stress circumferentially. These opposite principal stresses cause shearing planes to exist whose obliquity depends upon the relative magnitudes of the principal stresses by § 163. The scale breaks on these lines of shear and there result curves where the bright metal shows through, branching out from the hole, intersecting and fading away. The process of shearing a bar will develop the same curves from the flow of the metal on the face at the cut end. They are known as Cooper's lines. These lines show that deformation takes place at considerable distances from the immediate point of shearing or punching.

*Examples.*—1. A pull of 1,000 lb. per sq. in. and a thrust of 2,000 lb. per sq. in. are principal stresses. Find the kind, direction, and magnitude of the stress on a plane at 45° with either principal plane.

2. Find the stress per running unit of length of joint for a spiral riveted pipe when the line of rivets makes an angle of 45° with the axis of the pipe and when it makes an angle of 60°.   $0.707\ p$; $0.5\ p$.

3. A rivet is under the action of a shearing stress of 8,000 lb. per sq. in. and a tensile stress, due to the contraction of the rivet in cooling, of 6,000 lb. per sq. in. Find $p_1$ and $p_2$.
$$p_1 = -11,540 \text{ lb.}; \quad p_2 = +5,540 \text{ lb.}$$

4. A connecting plate to which several members are attached has a unit tension on a certain section of 6,500 lb. at an angle of 30° with the normal. On a plane at 60° with the first plane the unit stress of 5,000 lb. compression is found at 45° with its plane. Find the principal unit stresses and the shear.   $-6,600$; $+4,800$; $5,700$.

5. Assuming the weight of earth to be 105 lb. per cu. ft. and the horizontal pressure to be one-third the vertical, what is the direction and unit pressure per square foot on a plane making an angle of 15° with the vertical at a point 12 ft. under ground, if the surface is level?
515 lb.; $39\frac{1}{2}°$ with the horizon.

6. A stand-pipe, 25 ft. diam., 100 ft. high. The tension in lowest ring, if $\frac{7}{8}$ in. thick, is 7,440 lb. per sq. in. If plates range regularly from $\frac{7}{8}$ in. thick at base to $\frac{1}{4}$ in. at top, neglecting lap, the compression at base will be about 215 lb. per sq. in. For a wind pressure of 40

lb. per sq. ft., reduced 50% for cylindrical surface, and treated as if acting on a vertical section, $M$ at base $= 2,500,000$ ft.-lb. Compression on leeward side at base $= 485$ lb. per sq. in. If $p_1 = -7,440$ lb., $p_2 = 215 + 485 = 700$ lb., find the stress and its inclination for a plane at 30° to the vertical? $-6,193$ lb.; 34° 40'.

Prove that the shearing plane is 17° 04' from horizontal, and that the shear is 2,284 lb. per sq. in.

## CHAPTER XII.

### RIVETS: PINS.

**177. Riveted Joints.**—There are four different ways in which riveted joints and connections may fail. The rivets may shear off; the hole may elongate and the plate cripple in the line of stress; the plate may tear along a series of rivet-holes, more or less at right angles to the line of stress; or the metal may fracture between the rivet-hole and the edge of the plate in the line of stress. From the consideration that a perfect joint is one offering equal resistance to each of these modes of failure, the proper proportions for the various riveted connections are deduced.

**178. Resistance to Shear.**—The safe resistance of a rivet to shearing off depends upon the safe unit shear and the area of the rivet cross-section, which varies as the square of the diameter of the rivet. When one plate is drawn out from between two others, a rivet is sheared at two cross-sections at once, and is twice as effective in resisting any such action. Rivets so circumstanced are said to be in double shear, and their number is determined on that basis.

**179. Bearing Resistance.**—The resistance against elongation of the hole or crippling the plate depends on the safe unit compression and what is known as the *bearing* area, the thickness of the plate multiplied by the semicircumference of the hole. As the semicircumference varies as the diameter, it is more convenient, and sufficiently accurate, to use the product of the thickness of the plate and the diameter of the rivet with a value of allowable unit compression about fifty per cent greater than usual. In practice the bearing value is always given in terms of the diameter.

**180. Resistance of Plate.**—The resistance to tearing across the plate through a line of holes, or in a zigzag through two lines of holes in the same approximate direction, depends on the safe unit tensile stress multiplied by the cross-section of the plate after deducting the holes. If the transverse *pitch*, or distance between centres of rivets, is considerable, an assumption of uniform distribution of tension on that cross-section is not likely to be true.

The resistance of the metal between the rivet-hole and the edge of the plate in the line of stress is usually taken as the safe unit shear for the plate multiplied by the thickness and twice the distance from the rivet-hole to the edge. Some, however, consider that the resisting moment of the strip of metal in front of the rivet-holes is called into action.

**181. Bending: Friction.**—There are those who advise the computing of a rivet shank as if it were subjected to a bending moment. If the rivet fills the hole and is well driven, there is no bending moment exerted on it, unless it passes through several plates. As practical tests have shown that rivets cannot surely be made to fill the holes, if the combined thickness of the plates exceeds five diameters of the rivet, this limitation will diminish the importance of the question of bending.

No account is taken of the friction induced in the joint by its compression and the cooling of the rivet, and such friction gives added strength. As the rivet is closed up hot, the shank is under more or less tension when cold. Moreover, the head is not given the thickness required in the head of a bolt under tension, and therefore rivets are not available for any more tension, and should not be used for that purpose. Tight-fitting turned bolts are required in such a case.

**182. Spacing.**—The rivets should be well placed in a joint or connection, in order to insure a nearly uniform distribution of stress in the piece; they should be symmetrically arranged, be placed where they can be conveniently driven, and be spaced so that the holes can be definitely and easily located in laying out the work.

**183. Minimum Diameter of Rivets.**—The punch must have a little clearance in the die. The wad of metal shears out below the punch with more ease and with less effect on the surrounding metal when it can flow, as it were, a little laterally, and it then comes out as a smooth frustum of a cone with hollowed sides, reminding one of the *vena contracta*. The punch must also be a little larger than the rivet, to permit the ready entrance of the rivet-shank at a high heat. The diameter of the hole is commonly computed at $\frac{1}{8}$ inch in excess of the nominal diameter of the rivet; but the rivet is treated as if of its nominal diameter.

One other consideration has weight in determining the minimum diameter of the rivet. If the diameter of the rivet is less than the thickness of the plate, the punch will not be likely to endure the work of punching. A diameter one and a half times the thickness of the plate is often thought desirable.

**184. Number and Size of Rivets.**—Formulas are of little or no value in designing ordinary joints and connections. Boiler joints and similar work can be computed by formulas, but to no great advantage. Tables are used which give what is termed the shearing value of different rivet cross-sections in pounds for a certain allowable unit shear, and the bearing or compression value of different thicknesses of plate and diameters of rivet for a certain allowable unit compression. For a given thickness of plate, that diameter of rivet is the best whose two values, as above, most nearly agree. The quotient of the force to be transmitted through the connection or through a running foot of a boiler joint, divided by the less of the two practicable values, will give the minimum number of rivets. Their distribution is governed by the considerations previously referred to. Whether a joint in a boiler requires one, two, or three rows of rivets depends upon the number needed per foot.

*Example.*—Two tension bars, 6 in. by $\frac{1}{2}$ in., carrying 30,000 lb., are to be connected by a short plate on each side. Let unit shear be 7,500 lb. per sq. in., unit bearing 15,000 lb. on the diameter, and unit tension 12,000 lb. The bearing value of a $\frac{3}{4}$-in. rivet in a $\frac{1}{2}$-in. plate is 5,625 lb., its shearing value in double shear is $2 \times 3,310 = 6,620$ lb. Hence $30,000 \div 5,625 = 6$ rivets necessary. If these rivets can be so arranged that a deduction of but one rivet-hole is necessary from the cross-section of the tie, $(6 - (\frac{3}{4} + \frac{1}{8}))\frac{1}{2} = 2.56$ sq. in. net section, which will carry 30,750 lb. at 12,000 lb. unit tension. Each cover-plate cannot be less than $\frac{1}{4}$ in. thick, and, as will be seen presently, should be

RIVETS: PINS.                    195

made a little more. The length will depend on the distribution of the rivets, to be taken up next.

**185. Arrangement of Rivets.**—Long joints, under tension, like those of boilers, are connected by one or more rows of rivets,

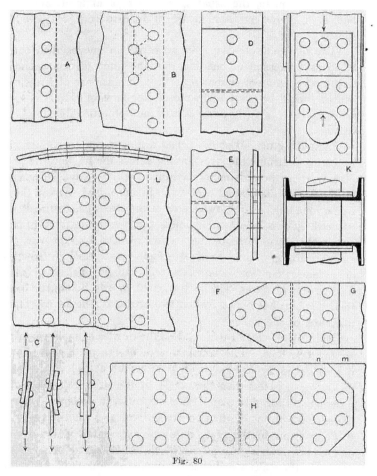

Fig. 80

as shown at A and B, Fig. 80. If more than one row is needed, the rivets are *staggered*, as at B, and the rows should be separated such a distance that fracture by tension is no more likely on a zigzag line than across a row. Experiments have shown that a plate will break along a zigzag line such as is shown at B unless the net length of that line exceeds by about one-third the net length of a line through a row. To prevent tearing out at the edge of the plate, the usual specification of at least one and a half rivet diameters from centre of hole to edge of plate will suffice.

The tendency of a lap-joint to cause an uneven distribution of stress by reason of bending, and the same tendency when a single cover is used, is shown at C. The increase of stress thus caused should be offset by increased thickness of plate. A cover strip on each side is preferable if not objectionable for other reasons.

In splicing ties, D shows a bad arrangement, the upper plan failing to distribute the stress evenly across the tie, and the lower plan wasting the section by excessive cutting away. The rivets at E are well distributed across the breadth, and weaken the tie by but one hole, as only two-thirds of the stress passes the section reduced by two holes; and, unless the net section at this place is less than two-thirds of the section reduced by one hole, it is equally strong. Thus $b-d=\frac{3}{2}(b-2d)$, or a breadth equal to or greater than four diameters will satisfy this requirement. The covers, however, will be weakened by two holes, and hence their combined thickness, when two are used, should exceed the thickness of the tie.

F similarly is better than G, and the tie at F is again weakened by but one hole. The sectional area of the plate shown at H is diminished by two holes at m and four at n, but the stress on section n is less than that on m by the stress the rivets at m transmit to the splice plate. Consequently, in designing a splice, if the area cut out by the two extra rivets at n multiplied by the working stress in the plate does not exceed the working value of the two rivets at m, the plate will be weakened by two holes only. In the splice shown the sections at m and n will be equally strong when $b-2d=\frac{14}{11}(b-4d)$ or when $b=18d$.

As it is desirable to transmit all but the proper fraction of the tension past the first rivet, the corners of the cover F or H are clipped off, thus increasing the unit tension in the reduced section and increasing its stretch to correspond more nearly with the unit tension and elongation of the tie beneath. The appearance of the connection is also improved.

It is not desirable to make splice plates as short as possible, because a short splice is likely to be weak. In splicing members built up of shapes this is especially the case, as a uniform distribution of stress over the whole cross-section is not easily secured when the stress passes into the member within a length less than its width. Short splice plates may be lengthened without increasing the number of rivets by using a greater pitch.

**186. Remarks.**—If the member is in compression, the holes are not deducted, since the rivets completely fill the holes; and the strength is computed on the gross section. Unless special care is exercised in bringing two connected compression pieces into close contact at their ends, good practice requires the use of a sufficient number of rivets at the connection to transmit the given force.

Rivet-heads in boiler work are flat cones. In bridge and structural work they are segments of spheres, known as button heads, and are finished neatly by means of a die. These heads may be flattened when room is wanting, and countersunk heads are used where it is necessary to have a finished flat surface.

Members of a truss which meet at an angle are connected by plates and rivets. The axes of the several members should if possible intersect in a common point. If they do not, moments are introduced which give rise to what are known as secondary stresses, as distinguished from the primary stresses due to the direct forces in the pieces of the frame. Such secondary stresses may be of considerable magnitude in an ill-designed joint.

It is desirable to arrange the rivets in rows which can be easily laid out in the shop, and to make the spacing regular, avoiding the use of awkward fractions as much as possible.

Commercial rivet diameters vary by eighths of an inch, $\frac{5}{8}$-, $\frac{3}{4}$-, and $\frac{7}{8}$-in. rivets being the ones frequently used. As much uniformity as possible in the size of rivets will tend to economy in cost.

**187. Structural Riveting.**—The following rules for structural work are in harmony with good practice:

Holes in steel $\frac{5}{8}$ in. thick or less may be punched; when steel of greater thickness is used, the holes shall be subpunched and reamed or drilled from the solid.

The diameter of the die shall not exceed that of the punch by more than $\frac{1}{16}$ of an inch, and all rivet-holes shall be so accurately spaced and punched that, when the several parts are assembled together, a rivet $\frac{1}{16}$ in. less in diameter than the hole can generally be entered hot into any hole.

The pitch of rivets, in the direction of the stress, shall never exceed 6 in., nor 16 times the thickness of the thinnest plate connected, and not more than 30 times that thickness at right angles to the stress.

At the ends of built compression members the pitch shall not exceed 4 diameters of the rivet for a length equal to twice the width of the member.

The distance from the edge of any piece to the centre of a rivet-hole must not be less than $1\frac{1}{2}$ times the diameter of the rivet, nor exceed 8 times the thickness of the plate; and the distance between centres of rivet-holes shall not be less than 3 diameters of the rivet.

The effective diameter of a driven rivet will be assumed to be the same as its diameter before driving; but the rivet-hole will be assumed to be one-eighth inch diameter greater than the undriven rivet.

In structural riveting these relationships between unit working stresses are very commonly used:

Bearing stress $= 1\frac{1}{2} \times$ tensile stress;

Shearing stress $= \frac{1}{2} \times$ bearing stress $= \frac{3}{4} \times$ tensile stress.

The shearing area of the rivets, therefore, should exceed by one-third the net area of the tension member they connect. See §§ 146, 147, 148.

**188. Boiler-riveting.**—Boiler work admits of standardization much more readily than structural work and standard boiler joints, which make the tensile strength of the net plate equal to the strength of the rivets, have been very generally adopted. A triple-riveted boiler joint is shown in Fig. 80, L. The most notable point of difference between boiler and structural riveting is that it is not customary to consider the bearing of rivets in boiler work. Rivets are figured for shear only and in double shear are considered to have but one and three-quarters time the value of rivets in single shear, instead of twice the value, as in structural work. Custom in this country makes the allowable unit shear on rivets approximately two-thirds the unit tensile stress in the plate, but the British Board of Trade rule gives about four-fifths.

The diameter of the rivets used should be about twice the thickness of the plate. In ordinary cases there is no danger that the rivets will be too far apart to render the joint water- or steam-tight, when the edge of the plate is properly closed down with a calking-tool.

**189. Pins: Reinforcing Plates.**—The pieces of a frame are frequently connected by pins instead of rivets. The axes of the several pieces are thus made to meet in a common point, if the pin-hole is central in each member. Pins are subjected to compression on their cylindrical surfaces, to shear on the cross-section, and to bending moments. The compression on the pin-hole is reduced to the proper unit stress, if necessary, by riveting reinforcing plates to the sides of the members, as shown at K, Fig. 80. A sufficient number of rivets to transmit the proper proportion of the force must be used, with a due consideration of the shearing value of a rivet and its bearing value in the reinforcing plate or the member itself, whichever gives the less value. No more rivets should be considered as efficient behind the pin than the section of the reinforcing plate each side of the pin-hole will be equivalent to.

When the pin passes through the web of a large built member, such as a post or a top chord of a bridge, the web is often so thin that more than one reinforcing plate on either side is needed. It is then economical to make the several plates of increasing length, the shortest on the outside, and determine the number of rivets in each portion accordingly.

Pin-plates should be made long for the same reason as given for making long splice plates. The longest plate is sometimes required to extend 6 in. inside the tie-plates so that the stress may be transferred to the flanges and not overtax the web. See K, Fig. 80.

200                STRUCTURAL MECHANICS.

**190. Shear and Bearing.**—The shear at any section of the pin is found from the given forces in the pieces connected. The resultant of the forces in the pieces on one side of any pin section will be the shear at that section. As the pin will probably not fit the hole tightly (a difference of diameter of one-fiftieth of an inch being usually permitted), the maximum unit shear will be four-thirds of the mean (§ 72). Specifications frequently give a reduced value for mean unit shear, which provides for this unequal distribution. Bearing area is figured as if projected on the diameter.

**191. Bending Moments on Pins.**—At a joint where several pieces are assembled, the resisting moment required to balance the maximum bending moment on the pin caused by the forces in those pieces will generally determine the diameter of the pin. In computing the bending moments, the centre line of each piece or bearing is considered the point of application of the force which it carries. This assumption is likely to give a result somewhat in excess of the truth, as any yielding tends to diminish the arm of each force.

The process of finding the bending moments will be made clear by an illustration. Fig. 81 shows the plan and elevation

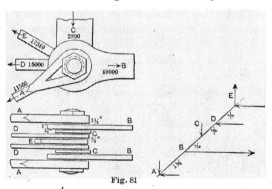

Fig. 81

of the pieces on a pin, with the forces and directions marked. The thickness of the pieces, which are supposed to be in contact, is also shown. The joint must be symmetrically arranged, to avoid torsion, and *simultaneous* forces must be used, which reduce to zero for equilibrium. As the joint is symmetrical, the computation is carried no farther than the piece adjoining the middle.

Resolve the given forces on two convenient rectangular axes, here horizontal and vertical. Set the horizontal components in order in the column marked $H$, the vertical ones in the column marked $V$. Their addition in succession gives the shears, marked $F$. The next column shows the distance from centre to centre of each piece. $Fdx$ is then the increment of bending moment; and the summation of increments gives, in the column $M$, the bending moment at the middle of each piece, from the horizontal and from the vertical components respectively. The square root of the sum of the squares of any pair of component bending moments will be the resultant bending moment at that section. It is comparatively easy to pick out the pair of components which will give a maximum bending moment on the pin. Equate this value with the resisting moment of a circular section and find the necessary diameter.

|   | $H.$ | $F.$ | $dx.$ | $Fdx.$ | $M.$ |
|---|---|---|---|---|---|
| A | +10,000 | +10,000 |  |  |  |
|   |  |  | $1\tfrac{1}{8}$ |  |  |
| B | −40,000 | −30,000 |  | +11,250 | +11,250 |
|   |  |  | $\tfrac{3}{4}$ |  |  |
| C | 0 | −30,000 |  | −22,500 | −11,250 |
|   |  |  | $\tfrac{5}{8}$ |  |  |
| D | +15,000 | −15,000 |  | −18,750 | −30,000 |
|   |  |  | $\tfrac{7}{8}$ |  |  |
| E | +15,000 |  |  | −13,125 | −43,125 |
|   | 0 |  |  |  |  |

|   | $V.$ |  |  |  |  |
|---|---|---|---|---|---|
| A | − 5,780 | − 5,780 |  |  |  |
|   |  |  | $1\tfrac{1}{8}$ |  |  |
| B |  | − 5,780 |  | − 6,503 | − 6,503 |
|   |  |  | $\tfrac{3}{4}$ |  |  |
| C | − 2,890 | − 8,670 |  | − 4,335 | −10,838 |
|   |  |  | $\tfrac{5}{8}$ |  |  |
| D |  | − 8,670 |  | − 5,419 | −16,257 |
|   |  |  | $\tfrac{7}{8}$ |  |  |
| E | + 8,670 |  |  | − 7,586 | −23,843 |
|   | 0 |  |  |  |  |

$M$ at $D = \sqrt{(30{,}000^2 + 16{,}257^2)}$; $M$ at $E = \sqrt{(43{,}125^2 + 23{,}843^2)}$. The latter is plainly the larger, and is 49,210 in.-lb.

The pieces can be rearranged on this pin to give a smaller moment. The maximum moment is not always found at the middle.

The bending moment at any point of the beam or shaft, when the forces do not lie in one plane, can be found in the same way.

A solution of the above problem by graphics can be found in the author's Graphics, Part II, Bridge Trusses.

*Examples.*—1. A tie-bar ½ in. thick and carrying 24,000 lb. is spliced with a butt-joint and two covers. If unit shear is 7,500 lb., unit bearing on diameter is 15,000 lb., and unit tension is 10,000 lb., find the number, pitch, and arrangement of ¾-in. rivets needed, and the width of the bar.

2. The longitudinal lap-joint of a boiler must resist 52,000 lb. tension per linear foot. If the unit working stress for the shell is 12,000 lb. and the other stresses as above, what size of rivet is best, for triple riveting, what the pitch, and the thickness of the shell?

# CHAPTER XIII.

## ENVELOPES.

**192. Stress in a Thin Cylinder.**—Boilers, tanks, and pipes under uniform internal normal pressure of $p$ per square inch.

Conceive a thin cylinder, of radius $r$, to be cut by any diametral plane, such as the one represented in Fig. 82, and consider

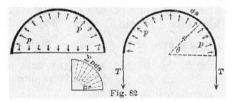

Fig. 82

the equilibrium of the half cylinder, which is illustrated on the left. It is evident that, for unity of distance along the cylinder, the total pressure on the diameter, $2pr$, must balance the sum of the components of the pressure on the semi-circumference in a direction perpendicular to the diameter. This pressure, $2pr$, uniformly distributed over the diameter, must cause a tension $T$ in the material at each end to hold the diameter in place. Hence

$$T = pr.$$

As all points of the circle are similarly situated, the tension in the ring at all points is constant and equal to $pr$. If the thickness is multiplied by the safe working tension $f$ per square inch, it may be equated with $pr$, giving

$$\text{Required net thickness} = pr \div f.$$

In a boiler or similar cylinder made up of plates an increase of thickness will be required to compensate for the rivet-holes. If $a$ is the pitch, or distance from centre to centre, of consecu-

tive rivets in one row along a joint, and $d$ the diameter of the *rivet-hole*, the effective length $a$ to carry the tension is reduced to $a-d$, and the *gross* thickness of plate must not be less than
$$\frac{pr}{f}\cdot\frac{a}{a-d}.$$

*Example.*—The circumferential tension in a boiler, 4 ft. diameter, carrying 120 lb. steam pressure is $120\cdot 24 = 2{,}880$ lb. per linear inch of length of shell, which will require a plate $\frac{2{,}880}{10{,}000}$ in. thick (net), if $f$ is not to exceed 10,000 lb. per sq. in. Net thickness $= \frac{9}{32}$ in. If a longitudinal joint has $\frac{3}{4}$-in. rivet-holes, at $2\frac{1}{4}$ in. pitch, in two rows, the thickness of plate must not be less than $\frac{2{,}880\cdot 2\frac{1}{4}}{10{,}000\cdot 1\frac{1}{2}} = \frac{7}{16}$ in.

**193. Another Proof** of the value of $T$ may be obtained as follows: The small force on arc $ds = p\,ds$. The vertical component of this force $= p\,ds\sin\theta = p\,dx$. The entire component on one side of the diameter is $\int_{-r}^{+r} p\,dx = 2pr$, which must be resisted by the tension in the material at the two ends of the diameter.

The same result will be obtained graphically by laying off a load line $=\Sigma p\,ds$, which becomes a regular polygon of an infinite number of sides, *i.e.*, a circle, with the lines to the pole making the radii of the length $pr$.

The cylinder, under these circumstances, is in stable equilibrium. If not perfectly circular, it tends to become so, small bending moments arising where deviation from the circle exists. Hence a lap-joint in the boiler shell causes a stress from the resisting moment to be combined with the tension at the joint.

The above investigation applies only to cylinders so thin that the tension may be considered as distributed uniformly over the section of the plate.

For riveting see Chapter XII.

**194. Stress in a Right Section.**—The total pressure from $p$ on a right section of the cylinder is $\pi r^2 p$, which will also be the resultant pressure on the head in the direction of the axis of the cylinder, whether the head is flat or not. This pressure causes tension in every longitudinal element of the cylinder, or in every cross-section. As this cross-section is $2\pi r \times$ thickness, the longitudinal tension per linear inch of a circumferential joint is

$\pi r^2 p \div 2\pi r = \tfrac{1}{2}pr$, or one-half the amount per linear inch of a longitudinal joint. Hence a boiler is twice as strong against rupture on a circumferential joint as on a longitudinal joint, and hence the longitudinal seams are often double riveted while the circumferential ones are single riveted.

**195. Stress in any Curved Ring under Normal Pressure.**—The stress in a circular ring of radius $r$, under internal or external normal unit pressure, $p$, is $pr$ per linear unit of section of the ring, being tension in the first case and compression in the second. Similarly, in single curved envelopes in equilibrium under normal pressure (that is, envelopes in which the stress acts in the direction of the shell) the stress at any point per linear unit along an element is equal to $p\rho$, in which $\rho$ is the radius of curvature of the curve cut out by a plane normal to the element and passing through the given point. Fig. 83 shows the trace of an arc of shell of width unity and of length $\rho d\theta$ acted upon from within by a normal pressure, $p$. Then $P = p\rho d\theta$ and for equilibrium the sum of the components acting in the direction of $P$ must be zero or $P = 2T \sin \tfrac{1}{2}d\theta = Td\theta$. Hence $T = p\rho$.

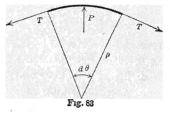

Fig. 83

**196. Thin Spherical Shell: Segmental Head.**—If a thin hollow sphere of radius $r'$ has a uniform normal unit pressure $p$ applied to it within, the total interior pressure on a meridian plane will be $\pi r'^2 p$, and the tension per *linear* inch of shell will be

$$\pi r'^2 p \div 2\pi r' = \tfrac{1}{2}pr'.$$

If $p$ is applied externally, the stress in the material will be compression. It may be noted that the *double* curvature of the sphere is associated with *half* the stress which is found in the cylinder of single curvature having the same radius.

If a segment of a sphere is used to close or cap the end of a cylinder or boiler, the same value will hold good. In this case the radius $r'$ is greater than $r$ for the cylinder.

If the segmental end is fastened to the cylinder by a bolted flange, the combined tension on the bolts will be $\pi r^2 p$, as this is the total force on a right section of the cylinder.

The flange itself will be in compression. The pressure $p$ from below, in Fig. 84, causes a pull per circumferential unit in the direction of a tangent at B, which pull has just been shown to be equal to $\tfrac{1}{2}pr'$. It may be resolved into vertical and horizontal components. The vertical component B C is, by § 194, $\tfrac{1}{2}pr$. The horizontal component $h$ must be proportioned to the vertical component as A O to A B, the sides of the right-angled triangle to which they are respectively perpendicular. As $AO = \sqrt{(r'^2 - r^2)}$,

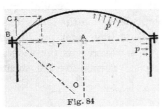
Fig. 84

$$h : \tfrac{1}{2}pr = \sqrt{(r'^2 - r^2)} : r,$$
or $$h = \tfrac{1}{2}p\sqrt{(r'^2 - r^2)}.$$

As $h$ is a uniform normal pressure applied from without (or tension applied from within) in the plane of the flange, the compression on the cross-section of the latter will be $hr$ or $\tfrac{1}{2}pr\sqrt{(r'^2-r^2)}$, to be divided by that cross-section for finding the unit compression.

Segmental bottoms of cylinders are sometimes turned inward. The principles are the same.

*Example.*—A segmental spherical top to a cylinder of 24 in. diameter, under 100 lb. steam pressure, has a radius of 15 in. with a versed sine of 6 in. The tension in top $= \tfrac{1}{2} \cdot 100 \cdot 15 = 750$ lb. per linear inch. If its thickness is $\tfrac{1}{4}$ in., the stress per sq. in. is 3,000 lb. The total pull on the flange bolts is $100 \cdot 144 \cdot 22 \div 7 = 45{,}260$ lb. A $\tfrac{3}{4}$-in. bolt has about 0.3 sq. in. section at bottom of thread, giving a tension value of about 3,000 lb. if $f = 10{,}000$ lb. There would be needed some 15 bolts, about $5\tfrac{1}{2}$ in. centre to centre on a circumference of 26 in. diameter. The compression in the flange is $\tfrac{1}{2} \cdot 100 \cdot 12 \cdot 9 = 5{,}400$ lb. A $2 \times \tfrac{1}{2}$-in. flange with a $\tfrac{3}{4}$-in. hole has a section $\tfrac{1}{2} \cdot 1\tfrac{1}{4} = \tfrac{5}{8}$ sq. in., giving a unit compression in the flange of $\tfrac{8}{5} \cdot 5{,}400 = 8{,}600$ lb. per sq. in.

A similar compression acts in the connecting circle between a water-tank and the conical or spherical bottom sometimes built. See §§ 204, 205.

**197. Thick Hollow Cylinder.**—If the walls of a hollow cylinder or sphere are comparatively thick, it will not be sufficiently accurate to assume that the stress in any section is uniformly distributed throughout it. If the material were perfectly rigid,

ENVELOPES.

the internal or external pressure would be resisted by the immediate *layer* against which the pressure was exerted, and the remainder of the material would be useless. As, however, the substance of which the wall is composed yields under the force applied, the pressure is transmitted from particle to particle, decreasing as it is transmitted, since each layer resists or neutralizes a portion of the normal pressure and undergoes extension or compression in so doing.

**198. Greater Pressure on Inside.**—Let Fig. 85 represent the right section of a thick hollow cylinder, such as that of a hydraulic press. Let $r_1$ and $r_2$ be the internal and external radii in inches; $p_1$ and $p_2$ the internal and external normal unit pressures in pounds per square inch, $p_1$ being the greater; and $p$ the unit normal pressure on any ring whose radius is $r$.

If a hoop is shrunk on to the cylinder, $p_2$ will be the unit normal pressure thus applied to the exterior of the cylinder.

The unit tensile stress found in a thin layer of radius $r$ and thickness $dr$ will be denoted by $t$, and will be due to *that portion of $p$ which is resisted by the layer and not transmitted to the next exterior layer.* The total tension on the radial section of a ring lying between $r_1$ and $r$ is $p_1 r_1 - pr$, since the pressure, $p_1$, on the inside sets up a tension of $p_1 r_1$ in the ring and the pressure, $p$, acting on the outside of the ring sets up a compression of $pr$. This total tension may also be expressed by $\int_{r_1}^{r} t\, dr$. As $p$ and $r$ are variables, there is obtained by differentiating the equation

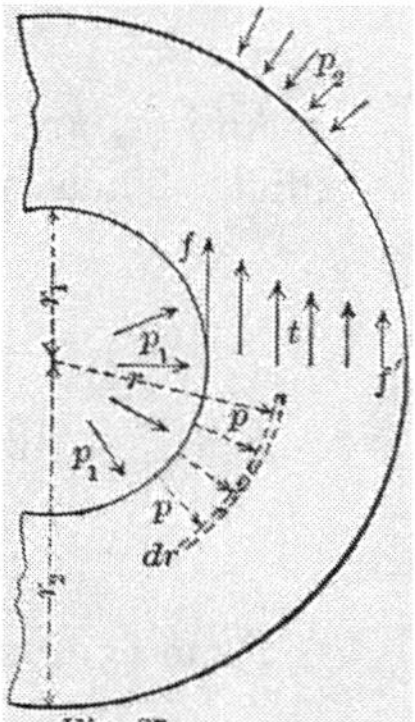

$$p_1 r_1 - pr = \int_{r_1}^{r} t\, dr,$$

$$-d(pr) = t\, dr,$$

or $\qquad p\, dr + r\, dp + t\, dr = 0. \quad . \quad . \quad . \quad (1)$

Another equation can be deduced from the enlargement of the cylinder. The fibres or layers between the limits $r_1$ and $r$, being compressed, will be diminished in thickness. The compression of a piece an inch in thickness by a unit stress $p$

Fig. 85

will be $p \div E$, § 10, and of one $dr$ thick will be $pdr \div E$. The total diminution of thickness between $r_1$ and $r$, from what it was at first, will therefore be $\dfrac{1}{E}\int_{r_1}^{r} p\,dr$.

But the annular fibre or ring whose radius is $r$ and length $2\pi r$ has been elongated $t \div E$ per inch of length. Its length will now be $2\pi r\left(1+\dfrac{t}{E}\right)$ and its radius $r\left(1+\dfrac{t}{E}\right)$. The internal radius must similarly have become $r_1\left(1+\dfrac{f}{E}\right)$, where $f$ is the value of $t$ for radius $r_1$. The thickness $r-r_1$ has now become $r\left(1+\dfrac{t}{E}\right) - r_1\left(1+\dfrac{f}{E}\right)$, and, by subtracting this value from $r-r_1$, there is found the diminution of thickness, $r_1\dfrac{f}{E}-r\dfrac{t}{E}$. This expression may be equated with the previous one for decrease of thickness, or

$$r_1\frac{f}{E}-r\frac{t}{E}=\frac{1}{E}\int_{r_1}^{r} p\,dr.$$

Since the first term is constant, there is now obtained by differentiating this equation,

$$-d(tr)=p\,dr, \quad \text{or} \quad t\,dr+r\,dt+p\,dr=0. \quad \ldots \quad (2)$$

Add (1) and (2), and multiply by $r$ to make a complete differential. Then integrate

$$2(t+p)r\,dr+r^2(dt+dp)=0;$$
$$r^2(t+p)=\text{constant}; \quad \therefore \ =r_1^2(f+p_1)=r_2^2(f'+p_2). \quad (3)$$

Again, subtract (1) from (2), and then integrate

$$dt-dp=0. \quad t-p=\text{constant}; \quad \therefore \ =f-p_1=f'-p_2. \quad (4)$$

From (3) and (4) are obtained, by addition and subtraction,

$$t=\frac{f-p_1}{2}+\frac{r_1^2}{r^2}\cdot\frac{f+p_1}{2}; \quad p=-\frac{f-p_1}{2}+\frac{r_1^2}{r^2}\cdot\frac{f+p_1}{2}. \quad (5)$$

## ENVELOPES.

If the internal radius is given, the external radius, and hence the required thickness, $r_2 - r_1$, is found by eliminating $f'$ from (3) and (4),

$$r_2 = r_1 \sqrt{\left(\frac{f + p_1}{f - p_1 + 2p_2}\right)}. \quad \ldots \ldots \quad (6)$$

If $p_2$ is atmospheric pressure, it may be neglected when $p_1$ is large. In that case

$$r_2 = r_1 \sqrt{\left(\frac{f + p_1}{f - p_1}\right)}; \quad \text{or} \quad r_2 - r_1 = r_1\left(\sqrt{\frac{f + p_1}{f - p_1}} - 1\right).$$

As $r_2$ becomes infinite when the denominator of (6) is zero, it appears that no thickness will suffice to bring $f$ within the safe unit stress, if $p_1$ exceeds $f + 2p_2$.

These formulas do not apply to bursting pressures, nor to those which bring $f$ above the elastic limit; for $E$ will not then be constant. They serve for designing or testing safe construction.

*Examples.*—Cylinders of the hydraulic jacks, for forcing forward the shield used in constructing the Port Huron tunnel, were of cast steel, 12 in. outside diameter, 8 in. diameter of piston, with ¼ in. clearance around same; pressure 2,000 lb. per sq. in.

$$\frac{r_2^2}{r_1^2} = \frac{36 \cdot 16}{289} = \frac{f + 2{,}000}{f - 2{,}000}. \quad f = 6{,}030.$$

A cast-iron water-pipe at the Comstock mine was 6 in. bore, 2½ in. thick, and was under a water pressure of 1,500 lb. on the sq. in., or about 3,400 ft. of water. Here $f = 2{,}770$ lb. per sq. in. for static pressure, while the formula for a thin cylinder gives 1,800 lb.

**199. Greater Pressure on Outside.**—In this case the direction or sign of $t$ will be reversed, it being compression in place of tension. From the preceding equations, without independent analysis, by making $t$ negative, there result:

$$-d(pr) = -t\,dr; \quad d(tr) = p\,dr.$$
$$p\,dr + r\,dp - t\,dr = 0; \quad t\,dr + r\,dt - p\,dr = 0.$$
$$r^2(p - t) = r_1^2(p_1 - f) = r_2^2(p_2 - f').$$
$$t + p = f + p_1 = f' + p_2.$$

The outer radius and pressure will now be taken as given quantities, and the unit compression in the ring at any point will be

$$t = \frac{f'+p_2}{2} + \frac{r_2^2}{r^2} \cdot \frac{f'-p_2}{2}; \quad p = \frac{f'+p_2}{2} - \frac{r_2^2}{r^2} \cdot \frac{f'-p_2}{2}. \quad (7)$$

$$r_1 = r_2 \sqrt{\left(\frac{f - 2p_2 + p_1}{f - p_1}\right)}, \quad \ldots \ldots (8)$$

which becomes, if $p_1$ is neglected as small,

$$r_1 = r_2 \sqrt{\left(1 - \frac{2p_2}{f}\right)}.$$

The external pressure $p_2$ must be less than $\frac{1}{2}(f+p_1)$, if $r_1$ is to have any value. It will be seen from $t$ in (7) that the compression is greatest at the interior.

*Example.*—An iron cylinder 3 ft. internal diameter resists 1,150 lb. per sq. in. external pressure. The required thickness, if $f = 9,000$ lb., is given by

$$18 = r_2 \sqrt{\left(1 - \frac{2,300}{9,000}\right)} = 0.86 r_2.$$

$r_2 = 20.9$ in. Thickness $= 3$ in.

**200. Action of Hoops.**—To counteract in a greater or less degree the unequal distribution of the tension in thick hollow cylinders for withstanding great internal pressures, hoops are shrunk on to the cylinders, sometimes one on another, so that before the internal pressure is applied, the internal cylinder is in a state of circumferential compression, and the exterior hoop in a state of tension. If the internal pressure on the hoop is computed, for a given value of $f$ in the hoop, and this pressure is then used for $p_2$ on the cylinder, the allowable internal pressure $p_1$ on the cylinder consistent with a permissible $f$ in this cylinder can be found. There is, however, an uncertainty as to the pressure $p_2$ exerted by the hoop.

## ENVELOPES. 211

*Examples.*—A hoop 1 in. thick is shrunk on a cylinder of 6 in. external radius and 3 in. internal radius, so that the maximum unit tension in the hoop is 10,000 lb. per sq. in. This stress, by § 198, will be due to an internal pressure on the hoop of 1,530 lb. per sq. in.

$$\text{For } 7 = 6\sqrt{\left(\frac{10,000 + p_1}{10,000 - p_1}\right)} \quad \text{or} \quad \frac{49}{36} = \frac{10,000 + p_1}{10,000 - p_1}.$$

This external pressure $p_2$ on the cylinder will cause a compressive unit stress in the interior circumference of the cylinder when empty, after the hoop is shrunk on, of 4,080 lb., and will permit an internal pressure in the bore of 8,448 lbs. per sq. in., consistent with $f = 10,000$ lb. For $\frac{36}{9} = \frac{10,000 + p_1}{10,000 - p_1 + 3,060}$. The cylinder alone, without the hoop, would allow a value of $p_1$ given by $\frac{36}{9} = \frac{10,000 + p_1}{10,000 - p_1}$, or $p_1 = 6,000$ lb. If the cylinder had been 4 in. thick, the internal pressure might have been 6,900 lb. The gain with the hoop, for the same quantity of material, is 1,548 lb., or some 22%.

Hydraulic cylinder for a canal lift at La Louviére, Belgium, 6 ft. 9 in. interior diameter, 4 in. thick, of cast iron, hooped with steel. Hoops 2 in. thick and continuous. When tested, before hooping, one burst with an internal pressure of 2,175 lb. per sq. in., one at 2,280 lb., and a third at 2,190 lb. These results, if the formula is supposed to apply at rupture, give an average tensile strength of 23,400 lb. per sq. in. The hoops were supposed to have such shrinkage that an internal pressure of 540 lb. per sq. in. would give a tension on the cast iron of 1,400 lb., and on the steel of 10,600 lb. per sq. in. The ram is 6 ft. 6¾ in. diam. and 3 in. thick, of cast iron, an example of the greater pressure outside.

**201. Thick Hollow Sphere.**—Greater pressure on inside. Let Fig. 85 represent a meridian section of the sphere. Suppose $f$, $t$, etc., to be perpendicular to the plane of the paper. The entire normal pressure on the circle of radius $r_1$ will be $p_1 \pi r_1^2$, and the tension on the ring between radii $r_1$ and $r$ will be $\pi(p_1 r_1^2 - p r^2)$. Any ring of radius $r$ and thickness $dr$ will carry $2\pi r t\, dr$, and hence is derived the first equation

$$\pi(p_1 r_1^2 - p r^2) = 2\pi \int_{r_1}^{r} r t\, dr, \quad \text{or} \quad -d(p r^2) = 2 r t\, dr.$$

$$\therefore r^2 dp + 2pr\, dr + 2rt\, dr = 0.$$

The second equation will be the same as obtained for the cylinder.

$$-d(tr) = p\, dr, \quad \text{or} \quad r\, dt + t\, dr + p\, dr = 0.$$

Strike out the common factor $r$ from the first equation, multiply the second by 2, and subtract.

$$2rdt - rdp = 0, \quad \text{or} \quad 2dt - dp = 0.$$

$$\therefore\ 2t - p = \text{constant}; \quad \therefore\ = 2f - p_1 = 2f' - p_2. \quad \cdot \ \cdot \quad (9)$$

Again, add the first to the second and multiply by $r^2$.

$$r^3(dp + dt) + 3r^2 dr(p + t) = 0.$$

$$\therefore\ r^3(p+t) = \text{constant}; \quad \therefore\ = r_1{}^3(f + p_1) = r_2{}^3(f' + p_2). \quad \cdot \quad (10)$$

From (9) and (10),

$$t = \frac{2f - p_1}{3} + \frac{r_1{}^3}{r^3} \cdot \frac{f + p_1}{3}; \quad p = -\frac{2f - p_1}{3} + 2\frac{r_1{}^3}{r^3} \cdot \frac{f + p_1}{3}.$$

$$r_2 = r_1 \sqrt[3]{\left(\frac{2(f + p_1)}{2f - p_1 + 3p_2}\right)}. \quad \cdot \ \cdot \ \cdot \ \cdot \ \cdot \quad (12)$$

These formulas are not applicable to bursting pressures for the reason given before. For a finite value of $r_2$, $p_1$ must be less than $2f + 3p_2$. If $p_2$ is atmospheric pressure, it may be neglected, and

$$r_2 = r_1 \sqrt[3]{\left(\frac{2(f + p_1)}{2f - p_1}\right)}.$$

**202. Sphere: Greater Pressure on Outside.**—Here again $t$ changes to compression or reverses in sign, yielding

$$t = \frac{2f' + p_2}{3} + \frac{r_2{}^3}{r^3} \cdot \frac{f' - p_2}{3}; \quad p = \frac{2f' + p_2}{3} - 2\frac{r_2{}^3}{r^3} \cdot \frac{f' - p_2}{3}.$$

$$r_1 = r_2 \sqrt[3]{\left(\frac{2f + p_1 - 3p_2}{2(f - p_1)}\right)}. \quad \cdot \ \cdot \ \cdot \ \cdot \ \cdot \quad (13)$$

That $r_1$ shall be greater than zero requires that $p_2 < \tfrac{1}{3}(2f + p_1)$.

**203. Diagrams of Stress.**—Curves may be drawn to represent the variation of $p$ and $t$ in the four preceding cases. They are all hyperbolic, and, if $r$ is laid off from the centre O on the horizontal axis, each curve will have the vertical axis through O for one asymptote, and for the other a line parallel to the horizontal axis, at a distance indicated by the first term in each value of $t$ or $p$. The four accompanying sketches show the various curves. The values of $f$ and $f'$, the unit stresses in the material at the interior and exterior, which correspond to the given values of $p_1$ and $p_2$, are found at the extremities of the abscissas which represent $r_1$ and $r_2$. The error which would arise from considering $t$ as uniformly distributed is manifest. The dotted

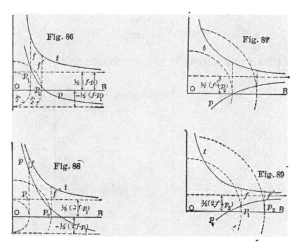

circles show the respective cylinders or spheres. Fig. 86 gives the external and internal tensile stress for $p_1$ in the interior of a thick cylinder. Fig. 87 shows the distribution of compression when the greater pressure is from without. Figs. 88 and 89 represent thick spheres under similar pressures.

**204. Tank with Conical Bottom.**—A water-tank of radius $r$ may be built with a conical bottom and be supported only at the perimeter by a circular girder. The pressure of the water in pounds per square inch at any point is $p = 0.434 \times$ depth of point below surface. In the cylinder the stress per unit of length on a vertical joint is $pr$; the stress on a horizontal joint is equal to the weight of the sides lying above the joint and is generally insignificant.

214   STRUCTURAL MECHANICS.

Any horizontal joint in the cone such as A A of Fig. 90 must carry a load, $W$, composed of the weight of water in the cylinder whose base is the circle A A, the water in the cone A B A and the weight of the metal shell of the same cone. The last item is comparatively insignificant. As the cone, like the sides, is built of thin plates, the stresses in the cone must always act tangentially to the shell, so $W \div 2\pi r_1$ is the vertical component of $T_1$, the stress per unit of length on the horizontal joint, and

$$T_1 = \frac{W \sec \theta}{2\pi r_1},$$

which varies from zero at B to a maximum at C.

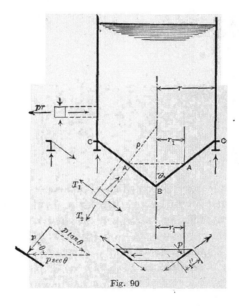

Fig. 90

To find the stress on a joint along an element of the cone imagine a ring of slant height unity to be cut out by two horizontal planes as shown in the figure. Substitute for the pressure, $p$, which acts normally around the ring, the two components, $p \tan \theta$ and $p \sec \theta$. Of these the former acting along the elements causes no stress on an elemental joint while the latter causes a tension in the ring of

$$T_2 = p r_1 \sec \theta = p\rho.$$

It can be proved that $r_1 \sec \theta$ is the radius of curvature of the conic section cut out by a plane normal to the element, B C, hence $T_2$ is equal to the pressure into the radius of curvature as shown in § 195. This tensile stress varies from zero at B to a maximum at C.

The total weight of the tank and contents is carried by a circular girder, which in turn is supported by three or more posts, consequently the girder is subjected to both bending and torsional moments. At C the stress, $T_1$, is resolved into vertical and horizontal components, the former of which is $W \div 2\pi r$ and is the vertical load per unit of length of girder; the latter is $\dfrac{W \tan \theta}{2\pi r}$ which causes a compressive ring stress in the girder of $\dfrac{W \tan \theta}{2\pi}$.

*Example.*—A circular tank, 40 ft. in diameter, has a conical bottom for which $\theta = 45°$. The depth of water above the apex is 60 ft. Weight of cu. ft. of water, 62.5 lb. Sec $\theta = 1.414$. Tension in lowest vertical ring of sides is $40 \times 0.434 \times 20 \times 12 = 4,170$ lb. per lin. in. Tension in radial joint of cone at A, half way up, is $50 \times 0.434 \times 120 \times 1.414 = 3,680$ lb. per lin. in. of joint. Same at C is $40 \times 0.434 \times 240 \times 1.414 = 5,900$ lb. per lin. in.

For tension on horizontal joint at A, half way up, $W = \pi \times 100 (50 + \tfrac{1}{3} \times 10) 62.5$. $T_1 = \tfrac{1}{2} \times 10 \times 53.33 \times 62.5 \times 1.414 = 23,600$ lb. per lin. ft. $= 1,960$ lb. per lin. in. of joint. At C, $T_1 = \tfrac{1}{2} \times 20 \times 46.67 \times 62.5 \times 1.414 = 41,250$ lb. per ft. $= 3,440$ lb. per lin. in. The vertical component of $T_1$ at C is $41,250 \div 1.414 = 29,200$ lb. per ft. of girder. As the horizontal and vertical components are equal, the compression in the girder is $29,200 \times 20 = 584,000$ lb.

**205. Tank with Spherical Bottom.**—The stresses in the spherical bottom are found in the same way as in a conical bottom. Any horizontal joint as A A carries the cylinder of water whose base is the circle A A, the segment of water A B A, and that part of the shell below the joint. The volume of the segment is $\pi a^2(r' - \tfrac{1}{3}a)$ if $a = r'(1 - \cos\theta)$. If the weight on the whole joint is $W$, the vertical component of the stress per linear unit of joint is $W \div 2\pi r' \sin\theta$ and the stress is

$$T_1 = \frac{W}{2\pi r' \sin^2\theta}.$$

The tension per unit of length on any meridian joint is, by § 196,

$$T_2 = \tfrac{1}{2} p r'$$

The vertical load on the girder is $W \div 2\pi r$ per unit of length, and the horizontal force applied to the girder by the bottom of the tank is $\dfrac{W}{2\pi r \tan\alpha}$ per unit of length, which causes a compressive stress of $\dfrac{W}{2\pi \tan\alpha}$ in the girder.

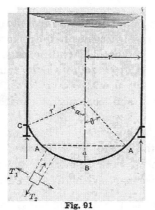

Fig. 91

The stresses in the spherical bottom are smaller than those in the conical bottom.

*Example.*—A circular tank, 40 ft. diameter and 40 ft. high, has a spherical bottom for which $\alpha = 45°$. Then $r' = 28.3$ ft. and the extreme height $= 48.3$ ft. Tension in radial joint at bottom $= \tfrac{1}{2} \times 62.5 \times 28.3 \times 48.3 = 42{,}645$ lb. per linear ft. $= 3{,}554$ lb. per in. of joint. Tension in radial joint at A, half way up, where $\theta = 22\tfrac{1}{2}°$, is 40,770 lb. per ft. or 3,397 lb. per in. of joint. At C, tension is 35,350 lb. per ft., or 2,946 lb. per in. of radial joint. Tension in horizontal joint at A is 41,760 lb. per ft., or 3,480 lb. per in., and at C is 39,220 lb. per ft., or 3,268 lb. per in. of joint. Compression in circular girder $= 554{,}700$ lb.

**206. Conical Piston.**—The cone C B C of Fig. 92 represents a conical piston of radius $r$, subtending an angle $2\theta$, with a normal steam pressure, $p$, per unit of area applied over its exterior or interior, the supporting force being supplied by the piston-rod at B. The force in the direction of the rod on any section A A of radius $r_1$ is $p\pi(r^2 - r_1^2)$ which becomes at the vertex $p\pi r^2$, the force on the piston-rod. This force will be compression on the rod and tension in the cone, if $p$ acts on the exterior of the cone,

ENVELOPES.

and the reverse if $p$ acts within the cone. The unit stress in the metal of the cone at this section will be found by multiplying this force by sec $\theta$, and then dividing by the cross-section, $2\pi r_1 t$, in which $t$ is the thickness of the metal. The unit stress on the circumferential section is then

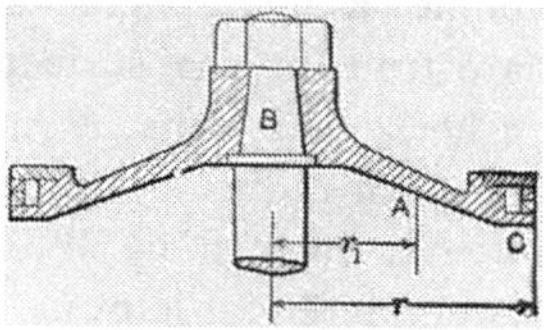

$$p_1 = \frac{p}{2r_1 t}(r^2 - r_1^2) \sec \theta,$$

Fig. 92

which is a maximum at the piston-rod if $t$ is constant.

The unit stress at A on a radial section is, by § 204,

$$p_2 = \frac{pp}{t} = \frac{pr_1}{t} \sec \theta.$$

When $p_1$ is compressive, $p_2$ is tensile and *vice versa*.

*Example.*—Conical piston, Fig. 92. $r = 24$ in., radius of rod = 3 in., $\theta = 69°$. Thickness for $r_1 = 17$ in. is 1.5 in.; for $r_1 = 8$ in. is 1.9 in. Steam pressure = 100 lb. per sq. in. Sec $\theta = 2.79$. For $r_1 = 17$ in., $p_1 = 100(24^2 - 17^2) 2.79 \div (2 \times 17 \times 1.5) = 1,570$ lb. per sq. in.; $p_2 = 100 \times 17 \times 2.79 \div 1.5 = 3,160$ lb. per sq. in. For $r_1 = 8$ in., $p_1 = 4,700$ lb. and $p_2 = 1,175$ lb. For alternating stresses on steel castings these stresses are satisfactory. See example § 175.

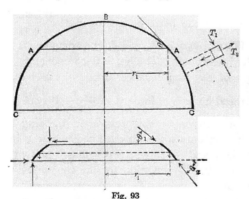

Fig. 93

**207. Dome.**—A dome subjected to vertical forces symmetrically placed around its axis, such as its own weight, may be treated as follows: C B C of Fig. 93 represents a meridian section of a dome, a hemisphere as shown, but the results to be deduced are true for any surface of revolution about a vertical axis. If a horizontal plane, A A, is passed through the dome to cut out a circle of radius $r_1$ and all the weight from the crown to that section is denoted by $W$, the stress in the shell per linear unit of circumferential joint is

$$T_1 = \frac{W \csc \theta}{2\pi r_1}.$$

$T_1$ is always compressive and is a maximum at the base.

To find the stress on a meridian section pass two horizontal planes through the dome so as to cut out a thin ring of mean radius $r_1$. If the total load above the ring is $W_1$ and the total vertical force supporting the ring is $W_2$, the weight of the ring is $W_2 - W_1$. As the shell of the dome is thin the stresses in the shell are tangential to the surface and the ring is acted upon by a system of forces around its circumference as shown on the right side of the figure. Resolve the forces into vertical and horizontal components as shown on the left. Acting upon the upper edge per unit of mean length of ring is the vertical component $\dfrac{W_1}{2\pi r_1}$ and the horizontal component $\dfrac{W_1 \ctn \theta_1}{2\pi r_1}$. By substituting $W_2$ and $\theta_2$ for $W_1$ and $\theta_1$ the components acting on the lower edge are found. The vertical components, together with the weight of the ring balance among themselves, but there is an unbalanced horizontal force of $H = \dfrac{1}{2\pi r_1}(W_1 \ctn \theta_1 - W_2 \ctn \theta_2)$ which causes tension or compression in the ring depending on whether it acts outward or inward. The stress in the ring is, therefore, $Hr_1$ and its intensity per linear unit of joint, $T_2$, is found by dividing by the width of the ring measured along the meridian. At the crown $T_2$ is compressive and equal to $T_1$, but it diminishes as A is taken lower and lower down and becomes tensile in the lower part of the structure.

**208. Resistance of a Ring to a Single Load.**—When a ring is acted upon by two equal and opposite forces as shown in Fig. 94, the curve becomes flatter at A and sharper at B, showing that

## ENVELOPES.

bending moments of opposite sign are set up at these points. From conditions of symmetry it is seen that each quadrant is acted upon by vertical forces of $\tfrac{1}{2}W$ together with an unknown moment at each end. Imagine the quadrant to be removed from

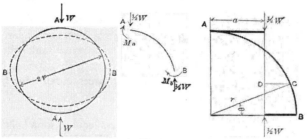

Fig. 94

the circle and horizontal levers to be attached at A and B so that the forces can be moved horizontally such a distance as to cause the actual moments existing at A and B. For equilibrium the forces, when moved, must be applied in the same line, and the moment at any section of the ring will be determined when the line of application of the forces is fixed.

To fix that line the deformation of the arc must be considered. By § 88, $\dfrac{1}{\rho}=\dfrac{M}{EI}$ which becomes $d\alpha=\dfrac{M}{EI}ds$ if the angle between the two radii of Fig. 44 is $d\alpha$. This equation gives the *change in the angle* between two right sections $ds$ apart, caused by the bending moment, and is true for curved beams as well as straight. To find the change between the two sections a distance $s$ apart integrate from zero to $s$. In the ring under consideration the tangents at A and B remain horizontal and vertical respectively, as seen from the condition of symmetry, hence the change in the angle between right sections at A and B is zero and as $E$ and $I$ are constant $\int_0^{\frac{1}{2}\pi r} M\,ds=0$. $M$ at any point C is $\tfrac{1}{2}W\times DC$ and must be expressed in terms of $r$ and $\theta$ to be integrated. Then

$$\tfrac{1}{2}Wr\int_0^{\frac{\pi}{2}}(a-r\cos\theta)d\theta=0;$$

$$\left[\int_0^{\frac{\pi}{2}} (a\theta - r\sin\theta) = 0 = \tfrac{1}{4}a\pi - r;\right.$$

$$a = \frac{2r}{\pi} = 0.6366r.$$

The bending moments in the hoop can now be determined.

If a ring is acted upon by four normal forces 90° apart as is the horizontal girder around the base of a water tank when the posts are inclined, the bending moments at various points of the girder may be found for each pair of forces independently and the results added algebraically.

*Examples.*—1. What is the net thickness required for a boiler shell 60 in. in diameter to carry 120 lb. steam pressure? What the gross thickness allowing for riveting, and the size and pitch of rivets? $\tfrac{3}{8}$ in.; $\tfrac{9}{16}$ in.; $\tfrac{3}{4}$ in.; 3 rows, 3 in. pitch.

2. What weight applied at top of circumference and resisted at bottom ought a cast-iron pipe, 12 in. diam., $\tfrac{1}{2}$ in. thick, and 6 ft. long, to safely carry, if $f = 12{,}000$ lb.?

3. Determine the thickness of the cast iron cylinder of a 10-ton hydraulic jack to work under a pressure of 1.000 lb. per sq. in. if $f = 4{,}000$ lb.

# CHAPTER XIV.

## PLATE GIRDERS.

**209. I Beam.**—A rolled beam of I section may be considered composed approximately of three rectangles,—two flanges, each of area $S_f$, and a web of area $S_w$. The depth between centres of stress of the flange sections may be denoted by $h'$, which is also very nearly the depth of the web. Then the resisting moment of the two flanges will be $fS_f h'$, and that of the web, since $M$ for a rectangle is $\tfrac{1}{6} f b h^2$, is $f \cdot \tfrac{1}{6} S_w h'$. The value for the entire section will be

$$M = f(S_f + \tfrac{1}{6} S_w) h'.$$

Hence comes the rule that one-sixth of the web may be added to one flange area in computing the resisting moment of an I beam. The extreme depth of the beam ought not, however, to be used for $h'$. The approximate distance between centres of gravity of the flanges will answer, since it is a little short of the true value for the flanges and a little longer than is correct for the web.

**210. Plate Girder.**—A portion of a plate girder and a section of the same is shown in Fig. 95. Such a structure acts as a beam and is designed to resist the maximum bending moments and shears to which it may be liable. It may be loaded on top, or through transverse beams connected to its web. It is used when the ordinary sizes of I beams are not strong enough to resist the maximum bending moment. As the flanges may be varied in section by the use of plates where needed, as shown at the right, there may be more economy of material in using a built beam rather than a rolled one, if the required maximum section is large.

222      STRUCTURAL MECHANICS.

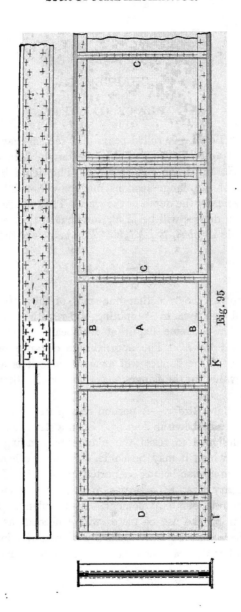

Fig. 95

The web A is made of sufficient section to resist the maximum shear, and the rest of the material is thrown into the flanges, B, where it will be farthest removed from the neutral axis and hence most efficient in resisting bending moment. As the thickness of the web plate is usually restricted to one-fourth inch, and in girders of any magnitude to three-eighths inch, as a *minimum*, it appears that the material in the web practically increases with the depth of the girder. As the stress in either flange multiplied by the distance between the centres of gravity of flanges resists the bending moment, the material in the flanges decreases as the depth increases; hence the most economical depth is that which makes the total material in the web, including the splice plates and stiffeners, as near as may be equal to that in the two flanges. Depth of beam contributes greatly to stiffness, when a small deflection is particularly desirable, and the depth may in such a case be so great as to make the web the heavier.

211. **Web.**—Some engineers apply the rule of § 209 to a plate girder and consider the flange section to be increased by *one-eighth* of the web section, the fraction one-eighth being used instead of one-sixth because of the weakening of the web by rivet-holes; but the more commonly received practice is to consider the flanges alone as resisting the bending moment at any section and the web as carrying all the shear uniformly distributed over its cross-section as shown in § 73. If the web is required to resist its share of the bending moment, web splices must be designed to transmit the stresses due to bending as well as those due to shear. Although web plates are generally made of the thinnest metal allowable, the designer is prevented from using too thin a web plate by the fact that thin plates offer so little bearing resistance to the rivets. If the plate is too thin it will generally be impossible to put in a sufficient number of rivets to connect the web to the flange.

212. **Flanges.**—If the maximum bending moments are computed for a number of points in the span of the girder, they can be divided by the allowable unit tensile stress times the effective depth (that is, the distance between centres of gravity of flanges) to give the required net sections of the tension flange at those

points. The compression flange is usually made of the same gross section as the tension flange. The deduction for rivet-holes in the latter, which is not necessary in the former, compensates for the slightly lower unit stress allowed for compression. The compression flange must be wide enough not to bend sideways like a strut between the points at which it is stayed laterally. As the web checks such lateral flexure in some degree, and as the flange stress varies from point to point, it is impracticable to apply a column formula. In practice the unsupported length of plate is not allowed to exceed a certain number of times the width of the flange: for railroad bridges different specifications give from 12 to 20; for girders carrying steady load only, as high a number as 30 may be used.

The selection of the plates and angles to give the required flange section is largely a matter of judgment. The angles must be large enough to support well the compression flange-plates and to be able to transmit the increments of stress from the web to such flange-plates. Hence the area of the flange-angles should be a considerable portion of the total flange section. For railway girders it is often specified that the section of the angles shall be at least one-half that of the whole flange, or that the largest angles procurable shall be used.

**213. Length of Flange-plates.**—As the bending moment varies from point to point in a girder, the required flange section will vary in the same way, being, in general, greatest at the half span. If the section at that point is made up of two angles with flange-plates, the girder may be made to approximate to a beam of uniform strength by running the flange-plates only such distance from the centre as they are needed.

Inspection of the necessary sections will show how far from the two ends of the girder, as I K, the flange-angles, with rivet-holes deducted, will suffice for the required flange section. From K to the corresponding distance from the other abutment the first plate must extend. A reasonable thickness being used for that plate, with a deduction for rivet-holes in the tension flange, it can again be seen where a second plate will be needed, if at all. This determination can be neatly made on a diagram of maxi-

PLATE GIRDERS.   225

mum moments, or on the similar diagram showing the necessary sections. Extend the plate either way a small additional distance, to relieve the angles and assure the distribution of stress to the plate. The thicker plate, if there is any difference in thickness, should be placed next to the angles.

When the girder carries a uniformly distributed load the

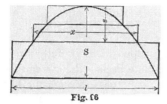 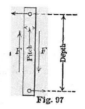

required flange areas vary as the ordinates to a parabola (Fig. 96) and the length of any flange-plate is given by

$$x = l\sqrt{\frac{s}{S}},$$

in which $l$ is the length of the girder from centre to centre of base-plates, $S$ the net flange area required at the centre, and $s$ the net area of the plate whose length is desired plus the area of such plates as may lie outside it.

If the girder is so long that the plates or angles must be spliced, additional cross-section must be supplied by covers at the splices, with lengths permitting sufficient rivets to transmit the force. Even compression joints, though milled and butted together, are spliced in good practice. The net area of the cover-plate and splice angles should be equal to that of the largest piece spliced. Only one piece should be cut at any one section, and enough lap should be given for the use of sufficient rivets to carry the stress the piece would have carried if uncut.

**214. Rivet Pitch.**—If a strip of web (Fig. 97) of a width equal to the pitch of the rivets connecting the flanges to the web is cut out by two imaginary planes, shearing forces, $F$, act on the two sides forming a couple with an arm equal to the pitch. Under the usual assumption that the web carries shear only, there are no other forces acting on the planes of section, and the only forces which can keep the strip from rotating are those supplied by the two flange rivets. Hence by equating the two couples the pitch can be found.

$$\text{Pitch} = \frac{\text{rivet value} \times \text{depth}}{\text{shear}}.$$

As the flange-angles are supposed to be fully stressed at the point where the first flange-plate begins, the increments of flange stress coming out of the web must pass through the flange-angles and into the flange-plate. The rivets connecting the flange-angles and the flange-plates must therefore resist the same stress as do the rivets connecting the web plate and the flange-angles, and the above formula applies to both, although the rivet values in the two cases will be different. The rivets through the web are in bearing (or double shear if the web is thick), while those in the flange-plate are in single shear and occur in pairs. But practically the pitch in one leg of an angle must be the same or an even multiple of the pitch in the other, so that the rivets may be staggered.

Make the pitch of rivets in inches and eighths, not decimals; do not vary the pitch frequently, and do not exceed a six-inch pitch, so that the parts may be kept in contact. If flange-plates are wide, and two or more are superimposed, another row of rivets on each side, with long pitch, may be required to insure contact at edges. Care must be taken that a local heavy load at any point on the flange does not bring more shear or bearing stress on rivets in the vertical legs of the flange-angles than allowed in combination with the existing stress from the web at that place and time.

Webs are occasionally doubled, making box girders, suitable for extremely heavy loads. The interior, if not then accessible for painting, should be thoroughly coated before assembling.

If the web must be spliced, use a splice plate on each side for that purpose, having the proper thickness for rivet bearing and enough rivets to carry all the shear at that section; there should be two rows of rivets on each side of the joint.

**215. Stiffeners.**—At points where a heavy load is concentrated on the girder, stiffeners, C, consisting of an angle on each side, should be riveted to throw the load into the web and to prevent the crushing of the girder. They should, for a similar reason, be used at both points of support, D. Such stiffeners act as columns and may be so figured, but as the stress in them varies from a maximum at one end to zero at the other, a good rule is to consider the length of the column equal to half the depth of the girder.

Since the thrust at 45° to the horizontal tends to buckle the web, and the equal tension at right angles to the thrust opposes the buckling, it is conceivable that a deep, thin web, while it has more ability to carry such thrust as a column or strut than it would have if the tension were not restraining it, may still buckle under the compressive stress; and it is a question whether stiffeners may not be needed to counteract such tendency. They might be placed in the line of thrust, sloping up at 45° from either abutment, but such an arrangement is never used. They are placed vertically, as at C, and spaced by a more or less arbitrary rule.

A common formula is: The web of the girder must be stiffened if the shear per square inch exceeds

$$10,000 - 75\frac{d}{t},$$

where $d=$ clear distance between flange-angles, or *between stiffeners* if needed, and $t=$ thickness of web. Another rule calls for stiffeners at distances apart not greater than the depth of the girder, when the thickness of the web is less than one-sixtieth of the unsupported distance between flange-angles. There is no rational method of determining the size of stiffeners used only to keep the web from buckling. The usual practice is about this: make the outstanding leg of the angle $3\frac{1}{2}$ inches for girders less than 4 feet deep, 4 inches for girders 5 feet deep, and 5 inches for girders over 7 feet deep.

Experience appears to show that stiffeners are not needed at such frequent intervals as the formula would demand. An insufficient allowance for the action of tension in the web in keeping the compression from buckling it, is probably the cause of the disagreement.

Interior stiffeners may be crimped at the ends, or fillers may be used under them to avoid the offset. End stiffeners and stiffeners under concentrated loads should not be crimped; they should fit tightly under the flange that the load may pass in at the ends.

*Example.*—A plate girder of 30 ft. span, load 3,000 lb. per ft., $f=15,000$ lb. per sq. in., $W=90,000$ lb., and $M_{max.}=\frac{1}{8}Wl$ $=4,050,000$ in.-lb. Assume extreme depth as 42 in., effective depth, 39 in. Net flange section at middle $=4,050,000 \div (39 \cdot 15,000) = 7$ sq. in. A $\frac{3}{8}$-in. web 42 in. deep will have $15\frac{3}{4}$ sq. in. area. Two flanges, each 7 sq. in. net+allowance for rivet-holes, will fairly equal the web. Use $\frac{3}{4}$-in. rivets.

Let the flange-angles be $2-4\times3\times\frac{3}{8}$ in. $=4.96$ sq. in. Deduct 2 holes, $\frac{3}{8}\times\frac{7}{8}=0.66$. Net plate $=7-4.3=2.7$ sq. in. A plate $9\times\frac{3}{8}=3\frac{3}{8}$ sq. in.; deduct two holes $=0.66$, leaving 2.71 sq. in. Two angles and plate, gross section $=4.96+3.37=8.33$ sq. in. Resisting moment of net section of angles $=4.3\times15,000\times3.25=209,625$ ft.-lb. Such a bending moment will be found at a distance $x$ from either end, given by $P_1x-\frac{1}{2}\cdot3,000x^2=209,625$. $x=5.9$ ft. $\therefore$ Cut off the plate 5 ft. from each end.

Shearing value of one $\frac{3}{4}$-in. rivet at 10,000 lb. per sq. in. $=4,400$ lb. Bearing value in $\frac{3}{8}$-in. plate at 20,000 lb. $=5,600$ lb. Max. shear in web $=45,000$ lb. Pitch for flange-angles, since bearing resistance is less than double shear, $=5,600\cdot39\div45,000=4.85=4\frac{3}{4}$ in. Make 3-in. pitch for 2 ft., then $4\frac{3}{4}$-in. for 6 ft., then 6-in. pitch to middle. Rivets in end web stiffeners, $45,000\div5,600=9$. Max. shear in web $=45,000$ $\div\{42\cdot\frac{3}{8}\}=2,780;\ 10,000-75\dfrac{d}{t}=2,800$, since $d=42-6$. No other stiffeners needed. By the other rule $36\div\frac{3}{8}=96$, and stiffeners are needed.

## CHAPTER XV.

### SPRINGS AND PLATES.

**216. Elliptic Spring.**—The elliptic spring is treated in § 97. If the load on the spring is $W$ and the span is $l$, the deflection is

$$v = \frac{1}{16} \frac{Wl^3}{EI_0}.$$

The fibre stress is

$$f = \frac{6M_0}{b_0 h^2} = \frac{3}{2} \frac{Wl}{b_0 h^2},$$

and the work done upon the spring is $\tfrac{1}{2}Wv$, which is equal to the resilience of the spring. Hence

$$\text{Resilience} = \frac{1}{32} \frac{W^2 l^3}{EI_0} = \frac{1}{6} \frac{f^2}{E} \cdot \text{volume}.$$

**217. Straight Spring.**—If a beam of uniform section, fixed at one end, has a couple or moment applied to it (Fig. 98) in place of a single transverse force, it will, as shown in § 89, bend to the arc of a circle. The deflection will be, if $l$ is the length of the beam,

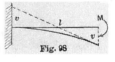

Fig. 98

$$v = \frac{Ml^2}{2EI} = \frac{fl^2}{2Ey_1}.$$

The work done by the rotation of a couple is the product of its moment by the angle through which it turns. For a deflection, $dv$, the free end of the spring turns through an angle $2dv \div l$; hence

$$\text{Resilience} = 2\int_0^v \frac{M}{l}dv = \frac{4EI}{l^3}\int_0^v v\,dv = \frac{2EI}{l^3}v^2 = \frac{ll}{2y_1^2}\cdot\frac{f^2}{E}.$$

For a rectangular section $bh$, these quantities become

$$v = \frac{6Ml^2}{Ebh^3} = \frac{fl^2}{Eh}.$$

$$\text{Resilience} = \frac{1}{6}\cdot\frac{f^2}{E}\cdot bhl = \frac{1}{6}\cdot\frac{f^2}{E}\cdot\text{volume}.$$

For a circular section the number 6 in the last expression will be replaced by 8.

**218. Coiled Spring.**—In practice the rectangular or cylindrical bar is bent into a spiral and subjected to a couple which, as a couple can be rotated in its plane without change, acts equally at all sections of the spring. The developed length of the spiral is $l$.

**219. Helical Spring.**—A cylindrical bar whose length is $l$ and diameter $d$, when fixed at one end and subjected to a twisting moment $T = Pa$ at the other, if the elastic limit is not exceeded, by § 84, is twisted through an angle $\theta = \frac{32Tl}{\pi Cd^4}$. The work expended in the torsion is

$$\tfrac{1}{2}T\theta = \frac{\pi Cd^4}{64l}\theta^2.$$

From § 84, $\theta = \frac{2q_1 l}{Cd}$, and therefore

$$\text{Resilience} = \frac{1}{4}\cdot\frac{q_1^2}{C}\cdot\frac{\pi d^2}{4}l = \frac{1}{4}\cdot\frac{q_1^2}{C}\cdot\text{volume}.$$

If $C = \frac{2}{5}E$ and $q_1 = \frac{4}{5}f$, work $= \frac{2}{5}\cdot\frac{f^2}{E}\cdot$ volume, while for flexure, as shown in the preceding section, work $= \frac{1}{6}\cdot\frac{f^2}{E}\cdot$ volume, a smaller

SPRINGS AND PLATES. 231

quantity; hence a spring of given weight can store more energy if the stresses are torsional than if they are bending.

If this bar is bent into a helix of radius, $a$, and the force, $P$, is applied at the centre in the direction of the axis of the cylinder, the moment, $Pa$, will twist the bar throughout its length. Then

$$q_1 = \frac{16Pa}{\pi d^3}, \quad \text{or} \quad P = \frac{\pi d^3}{16a}q_1.$$

The deflection of the spring is $v = a\theta$, since, as the force $P$ descends, the spring decends, and the action is the same as if the spring remained in place and the arm revolved through an angle $\theta$. The force $P$ is too small to cause any appreciable compression (or extension) of the material in the direction of its length.

$$v = a\theta = \frac{32}{\pi} \cdot \frac{Pa^2 l}{Cd^4} = \frac{2al}{d} \cdot \frac{q_1}{C} = \frac{4\pi n a^2}{d} \cdot \frac{q_1}{C},$$

if $n =$ number of turns of the helix and $l = 2\pi a n$.

If the section of the spring is not circular, substitute the proper value of $q_1$ or the resisting moment from § 85. If the rod is hollow, multiply the exterior volume by $\left(1 - \frac{d'^4}{d^4}\right)$. For a square section and a given deflection, $P$ will be about 65 per cent. of the load for an equal circular section. $C$ for steel is from 10,500,000 to 12,000,000.

*Example.*—A helical spring, of round steel rod, 1 in. diameter, making 8 turns of 3-in. radius, carries 1,000 lb.

$$q_1 = \frac{16 \cdot 1{,}000 \cdot 3 \cdot 7}{22} = 15{,}273. \quad v = \frac{4 \cdot 22 \cdot 8 \cdot 9 \cdot 15{,}273}{7 \cdot 1 \cdot 12{,}000{,}000} = 1.15 \text{ in.}$$

**220. Circular Plates.**—The analysis of plates supported or built in and restrained at their edges, and-loaded centrally or over the entire surface, is extremely difficult. The following formulas from Grashof's "Theorie der Elasticität und Festigkeit" may be used. The coefficient of lateral contraction is taken as $\frac{1}{4}$, or $m = 4$.

I. Circular plate of radius $r$ and thickness $t$, supported around its perimeter and loaded with $w$ per square inch.

$f_x =$ unit stress on extreme fibre in the direction of the radius, at a distance $x$ from the centre;

$f_y =$ unit stress perpendicular to the radius, in the plane of the plate, at the same distance $x$ from the centre.

$$f_x = \frac{45}{128}\frac{w}{t^2}(1\tfrac{3}{5}r^2 - 3x^2); \qquad f_y = \frac{45}{128}\frac{w}{t^2}(1\tfrac{3}{5}r^2 - x^2).$$

$$f_x = f_{y\ max.}\ (\text{for}\ x=0) = \frac{117}{128}\frac{wr^2}{t^2}. \qquad v_0 = \frac{189}{256}\frac{wr^4}{Et^3}.$$

For the same value of $t$, the maximum stress is independent of $r$, provided the total load $w\pi r^2$ is constant.

II Same plate, built in or fixed at the perimeter.

$$f_x = \frac{45}{128}\frac{w}{t^2}(r^2 - 3x^2); \qquad f_y = \frac{45}{128}\frac{w}{t^2}(r^2 - x^2).$$

At the centre, $f_x = f_y$. At the circumference $f_y$ is zero, and $f_x$ is maximum.

$$f_{x\ max.} = \frac{45}{64}\cdot\frac{wr^2}{t^2}. \qquad v_0 = \frac{45}{256}\frac{wr^4}{Et^3}.$$

III. Circular plate supported at the perimeter and carrying a single weight $W$ at the centre. Loaded portion has a radius $r_0$.

$$f_x = \frac{45}{32\pi}\frac{W}{t^2}\left(\log\frac{r}{x} - \frac{1}{5}\right); \qquad f_y = \frac{45}{32\pi}\frac{W}{t^2}\left(\log\frac{r}{x} + \frac{4}{5}\right).$$

These expressions become maxima for $x = r_0$, and the second is the greater.

$$v_0 = \frac{117}{64\pi}\frac{Wr^2}{Et^3}.$$

For values of $r \div r_0 =$ 10, 20, 30, 40, 50, 60,
$f_{max.} =$ 1.4  1.7  1.9  2.0  2.1  2.2. $W \div t^2$.

If $r_0=0$, the stress becomes infinite, as is to be expected, since $W$ will then be concentrated at a point, and the unit load becomes infinitely great. It is not well to make $r_0$ very small.

IV. Same plate, built in or fixed at the perimeter.

$$f_x = \frac{45}{32\pi} \frac{W}{t^2}\left(\log\frac{r}{x}-1\right); \quad f_y = \frac{45}{32\pi} \frac{W}{t^2}\log\frac{r}{x}.$$

The maximum value of $f$ is $f_y$, for $x=r_0$.

$$v_0 = \frac{45}{64\pi} \frac{Wr^2}{Et^3}.$$

For values of $r \div r_0 = $ 10, 20, 30, 40, 50, 60,
$f_{max.} = $ 1.0   1.3   1.5   1.6   1.7   1.8   $W \div t^2$.

**221. Rectangular Plates.**—The problem of the resistance of rectangular plates is more complex than that of circular plates. Grashof gives the following results:

V. Rectangular plate of length $a$, breadth $b$, and thickness $t$, $a > b$, built in or *fixed* at edges and carrying a uniform load of $w$ per square inch.

$$f_a = \frac{b^4 \cdot wa^2}{2(a^4+b^4)t^2}; \quad f_b = \frac{a^4 \cdot wb^2}{2(a^4+b^4)t^2}.$$

The most severe stress occurs at the centre in the direction $b$, that is, on a section parallel to $a$. If

$$a=b, \quad f = \frac{wa^2}{4t^2}.$$

The deflection at the centre is $v_0 = \frac{a^4 b^4}{a^4+b^4} \cdot \frac{w}{32Et^3}$, and for a square plate, $\frac{wa^4}{64Et^3}$.

VI. Plate carrying a uniform load of $w$ per square inch and supported at rows of points making squares of side $a$. Firebox sheet with staybolts.

$$f = \frac{15}{64} \frac{wa^2}{t^2}; \quad v_0 = \frac{15}{512} \frac{wa^4}{Et^3}.$$

Navier gives formulas for rectangular plates which are supposed to be very thin. Approximate values from those formulas are as follows:

VII. Rectangular plate, as in V, but *supported* around the edges.

$$f = 0.92 \frac{a^4 b^2}{(a^2+b^2)^2} \frac{w}{t^2}; \quad v_0 = 0.19 \frac{a^4 b^4}{(a^2+b^2)^2} \frac{w}{Et^3}.$$

VIII. Rectangular plate, supported at edges and carrying a single weight $W$ at centre.

$$f = 2.28 \frac{a^3 b}{(a^2+b^2)^2} \frac{W}{t^2}; \quad v_0 = 0.46 \frac{a^3 b^3}{(a^2+b^2)^2} \frac{W}{Et^3}.$$

For the same total load, $f$ is independent of the size of the plate, provided the ratio $a$ to $b$ and the thickness are unchanged.

*Example.*—A steel plate 36 in. square and $\frac{1}{4}$ in. thick, supported at edges, carries 430 lb. per sq. ft., or 3 lb. per sq. in. $f = 0.92 \cdot \frac{1}{4} \cdot 36 \cdot 36 \cdot 3 \cdot 16 = 14{,}300$ lb.

$$v = \frac{0.19}{4} \cdot \frac{36^4 \cdot 3 \cdot 4^3}{30{,}000{,}000} = \frac{1}{2} \text{ in.}$$

## CHAPTER XVI.

### REINFORCED CONCRETE.

**222. Reinforced Concrete.**—As concrete has small tensile strength and is likely to crack when built in large masses, it can be reinforced to advantage by steel bars or wire netting imbedded within it. This form of construction is much used and is especially applicable to beams and slabs, in which the steel is placed near the tension edge. By its use a great saving of material in masonry structures can often be effected, since the strength of the structure can be depended upon, rather than its weight. Among its advantages for buildings may be mentioned the fact that it is fireproof and that the metal is protected from rust. The expansion and contraction of the two materials from changes of temperature are so nearly alike that heat and cold produce no ill effects. When used in beams of any considerable span, there is the disadvantage that the dead load is a large proportion of the total load.

To compute the strength of a structure composed of two materials which act together, the modulus of elasticity of each material must be known, and it is here that the chief difficulty in computing the strength of reinforced concrete members lies. The modulus of elasticity of concrete is uncertain; it not only depends upon the composition of the concrete, but, for a given concrete, varies with the age, while for a particular specimen the stress deformation diagram is not a straight line as for steel, but the ratio of the stress to the deformation decreases with the load. The concrete takes a permanent set, even for small loads, in consequence of which a reinforced-concrete beam, after being released from a load, is in a condition of internal stress.

236                    STRUCTURAL MECHANICS.

Considering, then, our lack of knowledge of the exact stresses to be expected in the concrete, it is best to adopt as simple a method of computation as is reasonable.

**223. Beams.**—Beams may be figured according to the common theory of flexure, and the following assumptions will be made:

1. The steel and the surrounding concrete stretch equally.
2. Cross-sections, plane before bending, are plane after bending.
3. The modulus of elasticity is constant within the working stress.
4. The tension is borne entirely by the steel. As the tensile strength of concrete is low and as a crack in the beam would entirely prevent the concrete from resisting tension, this is a proper assumption to make and is on the side of safety.
5. There is no initial stress on the section.

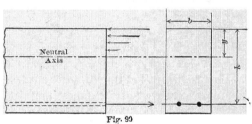

Fig. 99

From these assumptions it follows that the stresses acting on the cross-section are as shown in Fig. 99.

Let $E_s$ = modulus of elasticity of steel;

$E_c =$   "   "   "   " concrete;

$n = E_s \div E_c$;

$f_s$ = unit stress in steel;

$f_c =$   "   "   " concrete at extreme compression fibre;

$\lambda$ = deformation of fibre at unit distance from the neutral axis, between two sections originally unit distance apart;

$kbh$ = sectional area of steel;

$y$ = distance from neutral axis to extreme compression fibre;

$c = y \div h$;

$n$ = a numerical coefficient.

REINFORCED CONCRETE.          237

By § 10,
$$f_s = \lambda E_s(h-y), \qquad f_c = \lambda E_c y,$$

or the ratio between the unit stresses is

$$\frac{f_s}{f_c} = \frac{E_s}{E_c}\frac{h-y}{y} = e\frac{h-y}{y}. \qquad (1)$$

As the total tension on the cross-section must equal the total compression, for a rectangular cross-section

$$\tfrac{1}{2}byf_c = kbhf_s. \qquad (2)$$

Combining (1) and (2) gives

$$\frac{f_s}{f_c} = \frac{y}{2kh} = e\frac{h-y}{y}.$$

Solving this equation for $y$ gives the location of the neutral axis,

$$y = (-ek + \sqrt{e^2k^2 + 2ek})h = ch. \qquad (3)$$

This equation shows that for beams made of a given quality of concrete the location of the neutral axis depends only upon the percentage of reinforcement.

The resultant of the compressive stresses on the cross-section is applied at a point $\tfrac{1}{3}y$ from the top of the beam and together with the tensile stress in the steel forms a couple whose arm is $h - \tfrac{1}{3}y$. The moment of this couple, which is the resisting moment of the beam, is

$$M = \tfrac{1}{2}byf_c(h - \tfrac{1}{3}y) \qquad (4)$$
$$= kbhf_s(h - \tfrac{1}{3}y),$$

the first expression giving the moment in terms of the stress in the concrete and the second in terms of the stress in the steel. If the value of $y$ from (3) is substituted in (4) there results

$$M = \tfrac{1}{2}c(1 - \tfrac{1}{3}c)f_cbh^2 = n_cf_cbh^2 \qquad (5)$$
$$= k(1 - \tfrac{1}{3}c)f_sbh^2 = n_sf_sbh^2.$$

238   STRUCTURAL MECHANICS.

As long as $f_c$ and $f_s$ are the unit stresses in the concrete and in the steel, the two forms of (5) must be equal, but if $f_c$ and $f_s$ are arbitrarily chosen working stresses, that form of (5) must be used in designing which will give the smaller value of $M$. Therefore use $n_c$ when $n_c f_c < n_s f_s$ or when $\dfrac{f_s}{f_c} > \dfrac{c}{2k}$. The actual fibre stresses, however, can be made to assume any given ratio by changing the proportion of steel reinforcement, and that proportion can be found by eliminating $y$ from (1) and (2) with the result

$$k = \frac{1}{2}\frac{ef_c^2}{f_s^2 + ef_cf_s} = \frac{e}{2\dfrac{f_s}{f_c}\left(\dfrac{f_s}{f_c}+e\right)}. \quad \ldots \ldots (6)$$

Equation (5) can be readily applied to the design of beams by tabulating the values of $n_c$ and $n_s$ for different values of $k$.

| $k$ | $e=8$ | | | $e=10$ | | | $e=12$ | | |
|---|---|---|---|---|---|---|---|---|---|
| | $c$ | $n_c$ | $n_s$ | $c$ | $n_c$ | $n_s$ | $c$ | $n_c$ | $n_s$ |
| .003 | .196 | .092 | .00280 | .217 | .101 | .00278 | .235 | .108 | .00277 |
| .004 | .223 | .103 | .00370 | .246 | .113 | .00367 | .265 | .121 | .00368 |
| .005 | .246 | .113 | .00459 | .270 | .123 | .00455 | .291 | .131 | .00452 |
| .006 | .266 | .121 | .00546 | .292 | .132 | .00542 | .314 | .140 | .00537 |
| .007 | .284 | .128 | .00633 | .311 | .140 | .00628 | .334 | .148 | .00622 |
| .008 | .300 | .135 | .00720 | .328 | .146 | .00713 | .353 | .155 | .00706 |
| .010 | .328 | .146 | .00891 | .358 | .157 | .00880 | .384 | .167 | .00872 |
| .012 | .353 | .155 | | .384 | .167 | | .411 | .177 | |
| .014 | .375 | .164 | | .407 | .176 | | .435 | .186 | |
| .016 | .394 | .171 | | .428 | .183 | | .457 | .194 | |
| .018 | .411 | .177 | | .447 | .190 | | .476 | .200 | |
| .020 | .428 | .183 | | .463 | .196 | | .493 | .206 | |

When reinforced concrete beams are tested to destruction, they sometimes fail by the opening of diagonal cracks which seem to follow in a general way the lines of principal stress. To prevent that mode of failure most designers provide some form of reinforcement in a vertical plane. Stirrups or loops of wire lying in planes of section and spaced at intervals less than the depth of the beam are often employed, but no satisfactory method of determining their size and spacing has been proposed.

Another mode of failure to be guarded against is the slipping of the steel in the surrounding concrete. That the beam may not be weak in this respect, the change of stress in the bars, between two sections a short distance apart (say one inch), must not exceed the area of the surface of the bars between the two sections multiplied by the safe unit adhesive stress of concrete to steel. This change of stress between two sections one inch apart is $F \div h(1-\frac{1}{3}c)$, since the shear in the beam measures the change of bending moment at any section, as is shown in § 56. Rods which have been roughened or corrugated may be used to diminish the likelihood of slipping.

*Example.*—Design a beam of 16 ft. span to carry an external load of 500 lb. per ft. if $f_c = 500$ lb., $f_s = 16,000$ lb., and $e = 10$, assuming the beam to weigh 300 lb. per ft. $M = \frac{1}{8} \cdot 800 \cdot 16 \cdot 16 \cdot 12 = 307,200$ in.-lb. If each material is to be stressed to its limit, the proportion of reinforcement needed is $10 \div 2 \cdot 32 \cdot 42 = 0.00372$, or say 0.4%. Then $n_c = 0.113$ and $bh^2 = 307,200 \div (0.113 \times 500) = 5,430$. $b = 12\frac{1}{4}$ in., $h = 21$ in., and the area of steel is $0.004 \times 12\frac{1}{4} \times 21 = 1.03$ sq. in., or say $4 - \frac{9}{16}$ in. rounds. If there are 2 in. of concrete below the rods, the beam weighs $150 \times 12\frac{1}{4} \times 23 \div 144 = 284$ lb. per ft.

If 0.8% of steel is used, instead of 0.4%, $bh^2 = 4210$. $b = 10\frac{1}{2}$ in., $h = 20$ in., and 1.68 sq. in. of steel is needed; use $1 - \frac{3}{4}$ in. and $2 - \frac{7}{8}$ in. rounds.

Adhesion between concrete and steel. $F_{max.} = 800 \times 8 = 6,400$ lb. For first beam $6,400 \div 21(1 - \frac{1}{3} \times 0.246) = 332$ lb. Superficial area of $4 - \frac{9}{16}$ rounds $= 4 \cdot \frac{9}{16} \cdot \frac{22}{7} = 7.07$ sq. in. $332 \div 7.07 = 47$ lb. per sq. in. For second beam $359 \div 7.86 = 46$ lb. per sq. in.

**224. Columns.**—Reinforced concrete columns are built with steel rods embedded parallel to and spaced symmetrically about the axis. The rods should be tied together by wire or bands at intervals not greater than the diameter of the column. A common design is a square section with the rods near the corners. As the ratio of length to breadth is generally small, it is usual to design such columns as short blocks.

The ratio between the intensity of the stress in the steel and in the concrete is

$$\frac{f_s}{f_c} = \frac{E_s}{E_c} = e,$$

and the total load the column may carry is

$$P = f_c S(1-k) + f_s kS = f_c S[1 + k(e-1)],$$

in which $S$ is the total area of cross-section of the column.

With the usual working stresses and ratio of $E_s$ to $E_c$ employed in designing, the greatest allowed unit stress in the reinforcement is so small that economy requires the amount of steel to be reduced as much as reasonable, hence the load $P$ will be but slightly greater than $f_c S$.

**225. Safe Working Stresses.**—The following safe unit stresses may be used in buildings:

| | |
|---|---|
| Concrete, bending........................ | 500 lb. per sq. in. |
| "       direct compression............... | 350 "   "   "   " |
| Steel, tension........................... | 16,000 "   "   "   " |
| Adhesion of concrete to steel............. | 50 "   "   "   " |

$E_s \div E_c = 10$ is a fair average value; the ratio may vary considerably without materially affecting the proportions of the beam.

The weight of reinforced concrete is about 150 lb. per cu. ft.

The concrete should be rich; one part cement, two, sand; four, broken stone is a good mixture, the stone being broken to pass through a ¾-inch ring. It should be mixed wet and placed with great care to insure the proper bedding of the steel.

The proportion of steel reinforcement to use in beams is a question of economy, which is most easily solved by designing a number of sections of the same ratio of $b$ to $h$, but with different values of $k$, and figuring the cost per foot of each. A considerable variation in $k$ affects the cost but slightly, for ordinary prices of concrete and of steel. The percentage of reinforcement is usually between ½ and 1½; $k = 0.007$ is an average value.

The concrete lying below the reinforcing rods serves merely to protect the steel. For fireproofing two inches is sufficient. Steel thoroughly covered with concrete does not rust, and bars, which were covered with rust when placed in concrete, have been found to be bright when removed after some time.

# INDEX.

Angle of repose, 181
Annealing, 29
Ashlar, 30
Axis, neutral, 56, 61, 79

Bauschinger's experiments, 154
Beams, 2, 51, 87, 107
  bending moments, 41, 44
  cantilever, 45
  Clapeyron's formula, 118
  column and beam, 149
  continuous, 116
  curved, 60
  deflection, 87
  elastic curve, 87
  fixed, 107
  flexure of, 87
  flitched, 102
  I beams, 69, 173, 221
  impact, 103
  inclined, 60
  modulus of rupture, 59
  moving loads on, 48
  neutral axis, 56
  oblique loading, 78
  reactions, 38
  reinforced concrete, 236
  resilience of, 103
  restrained, 107
  sandwich, 102
  shaft and beam, 84
  shear, external, 43
  shear, internal, 65
  slope, 88
  stiffness, 89
  stresses in, 54, 65, 179
  three-moment theorem, 116, 122
  tie and beam, 130
  timber, 68
  torsion on, 84
  uniform strength, 63, 97
  work, internal, 104

Bending and compression, 149
  and tension, 130
  and torsion, 84
Bending moment, see Moment, 41
Bessemer process, 25
Blocks in compression, 15, 137
Boilers, 189, 203
  rivets, 198
  working stresses, 166
Bricks, 31
Bridges, shear in panel, 52
  working stresses, 162, 164
Buildings, working stresses, 165
Burnettizing, 21

Cantilevers, 45, 47
Carbon and iron, 22
Cast iron, 22
  properties of, 23
  working stresses, 166
Cement, 33
Centrifugal force, 133
Clapeyron's formula, 118
Clay, 31
Coefficient, see Modulus.
Columns, 2, 137
  beam and column, 149
  deflection, 142
  designing, 146
  eccentric load, 148
  ends, fixed or hinged, 146
  Euler's formula, 142
  flexure, direction of, 141, 146
  Gordon's formula, 144
  ideal column, 142
  lacing-bars, 151
  pin ends, 147
  radius of gyration, 146
  Rankine's formula, 144
  reinforced concrete, 239
  short, 137, 145
  straight-line formula, 147

# INDEX.

Columns, swelled, 148
    timber, 162
- transverse force on, 149
    working stresses, 162
    yield-point, 141
Combined stresses, 3
    bending and compression, 149
    " " tension, 130
    " " torsion, 84
    tension and torsion 133
Compression, 3, 15, 137
    bending and, 149
    eccentric load, 137
    granular materials under, 15
Concrete, 35, 235
Concrete-steel, 235
Cone, stresses in, 213, 216
Connecting-rod, 132
Continuous beams, 116
Cooper's lines, 190
Crank, 84
Curvature of beams, 88
Curve, elastic, 87
    stress-deformation, 9
Cylinders, thick, 206
    thin, 203

Deflection of beams, 87
    simple, 89
    restrained, 107
    uniform strength, 97
Deflection of columns, 142
Deformation, 6, 184
Distortion, 7, 186
Dome, 217
Ductility, 6, 17, 26, 29

Earth pressure, 181
Eccentric load, 127, 137, 148
Elastic curve, 87
Elastic limit, 9, 154
Elasticity, modulus of, $E$, 6
    cast iron, 23
    concrete, 235, 240
    steel, 26
    stone, 30
    timber, 22
    wrought iron, 24
Elasticity, shearing modulus of, $C$, 7, 186
Ellipse of stress, 174
Elongation, work of, 8, 10, 13, 27
Envelopes, 203
Equilibrium, conditions of, 1
Euler's formula, 142
Eyebars, 131, 134

Fatigue of metals, 154
Flexure, common theory of, 54

Girders, see Beams.
Girder, plate, see Plate girder, 221
Gyration, radius of, 71

Hooks, 129
Hoops, 210

I beam, 69, 173, 221
Impact, 159
Inertia, moment of, 57, 71
    product of, 76
Iron, cast, 22
    malleable, 28
    wrought, 24

Joints, masonry, 138
    riveted, 195

Lattice bars, 151
Launhardt-Weyrauch formula, 156
Lime, 32
Lines of principal stress, 180
Linseed oil, 36
Loads, dead and live, 159
    eccentric, 127, 137, 148
    sudden application, 14
    wheel loads on beam, 49

Machinery, working stresses, 166
Manganese in steel, 26
Masonry, 30
    working stresses, 166
Materials, 19
Middle third, 138
Modulus of elasticity, see Elasticity, 6
    of resilience, 13
    of rigidity, 186
    of rupture, 59
    section, 57
Moment, bending, 41
    maximum, 44
    on pins, 200
    position of load for maximum, 49
    sign of, 41
Moment of inertia, 71
Moment of resistance, 57
    oblique loading, 78
Moment, torsional, 81
Mortar, 32

Neutral axis, 56, 61, 79
Nuts, 134

Oblique load on beam, 78
Open-hearth process, 25

Paint, 36
Parallel rod, 133
Pedestals, 166
Permanent set, 8

# INDEX.

Phosphorus in steel, 26
Pier-moment coefficients, 121
Pig iron, 23
Pins, 109, 200
  distribution of shear in, 67
  friction, 147
  working stresses, 163
Pipes, 203
Piston, conical, 189, 216
Plaster, 33
Plate girder, 221
  stresses in web, 69, 173
Plates, resistance of, 231
Polar moment of inertia, 71
Portland cement, 34
Posts, see Columns, 137
Power, shaft to transmit, 83
Principal stresses, 170
  lines of, 180
Product of inertia, 76
Puddling furnace, 24
Pull and thrust at right angles, 172
Punching steel, effect of, 28

Radius of gyration, 71
Rafter, 60
Rankine's column formula, 144
Rankine's theory of earth pressure, 181
Reactions of beams, 38
Rectangular beams, 58
Repose, angle of, 181
Resilience, definition, 13
  modulus of, 13, 103
  of bar, 13
  of beam, 103
  of springs, 229
Resisting moment, see Moment, 57
Restrained beams, 107
Retaining wall, 182
Rigidity, modulus of, 187
Ring under normal pressure, 205
  under single load, 218
Rivets, 192
  plate girder, 225
  steel for, 26
  working stresses, 163
Rollers, 163
Rubble, 31
Rupture, modulus of, 59

Safe working stresses, 153
Sandwich beams, 102
Screw threads, 134
Secondary stresses, 158, 197
Section modulus, 57
Set, permanent, 8
Setting of cement, 34
Shafts, 81
  working stresses, 166
Shear, 3, 43

Shear, deflection due to, 105
  derivative of bending moment, 45
  distribution on section of beam, 65
  modulus of elasticity, $C$, 186
  position of load for maximum, 48, 50
  sign of, 43
  timber beams, 68
  two shears at right angles, 169, 172, 186
  work of, 105
Shearing planes, 176
Sign of bending moment, 41
  compression and tension, 4
  shear, 43
Silicon in iron, 23
Slope of beam, 88
Spangenberg's experiments, 153
Sphere, stresses in, 205, 211
Splices, 195
  in plate girder, 223, 225, 226
Springs, 98, 229
Steel, 25
  shearing and punching, 28
  structural, 26
  tool, 28
  working stresses, 162
Steel concrete, 235
Stiffness of beams, 89
Stiffeners, 227
Stirrups, 238
Stone, 29
Straight-line formula, 147
Strain, see Deformation.
Strength, beams of uniform, 63, 97
  cross-sections of equal, 62
  ultimate, 11
Strength of cast iron, 23
  steel, 26
  timber, 22, 160
  wrought iron, 24
Stresses, 2, 167
  alternating, 154
  conjugate, 170
  distribution on section of beam, 54, 65, 179
  ellipse of, 174
  internal, 2, 167
  lines of principal, 180
  principal, 170
  reversal of, 155, 162
  secondary, 158, 197
  sign of, 3
  unit, 4, 167
  working, 160, 240
Stress-deformation diagram, 9
Struts, see Columns, 2, 137
Sudden loading, effect of, 14
Sulphur in steel, 26

Tanks, 213

# INDEX.

Tempering, 26, 28
Tension, 3, 127
  bending and, 130
  connections, 133, 196
  eccentric load, 127
  torsion and, 128, 133
Three-moment theorem, 116, **122**
Tie, 2, 127
  and beam, 130
Timber, 19
  column formulas, 161
  modulus of elasticity, 22
  shear in beams, 68
  strength of, 22
  working stresses, 160
Torsion, 81
  bending and, 84
  resilience of, 230
  tension and, 133
  twist of shaft, 83
Trees, growth of, 19
Twist of shaft, 83

Ultimate strength, 11

Uniform strength, beams of, 63, 97
Unit stresses, 4

Varnish, 36
Varying cross-section, 12
Volume, change of, 183

Wall, retaining, 182
  middle third, 138
Web of plate girder, 223
  stresses in, 69, 173
Welding, 26
Wheel loads, 49
Wöhler's experiments, 153
Wood, 19
Work of elongation, 8, 10, 13, 27
  flexure, 104
  shear, 105
  springs, 229
Working stresses, 153
  reinforced concrete, 240
Wrought iron, 24

Yield-point, 10

# SHORT-TITLE CATALOGUE

OF THE

## PUBLICATIONS

OF

## JOHN WILEY & SONS,

NEW YORK.

LONDON: CHAPMAN & HALL, LIMITED.

ARRANGED UNDER SUBJECTS.

Descriptive circulars sent on application. Books marked with an asterisk (*) are sold at *net* prices only. All books are bound in cloth unless otherwise stated.

### AGRICULTURE—HORTICULTURE—FORESTRY.

| | | |
|---|---|---|
| Armsby's Manual of Cattle-feeding................................12mo, | $1 | 75 |
| Principles of Animal Nutrition...........................8vo, | 4 | 00 |
| Budd and Hansen's American Horticultural Manual: | | |
| Part I. Propagation, Culture, and Improvement................12mo, | 1 | 50 |
| Part II. Systematic Pomology...................................12mo, | 1 | 50 |
| Elliott's Engineering for Land Drainage..........................12mo, | 1 | 50 |
| Practical Farm Drainage.....................................12mo, | 1 | 00 |
| Graves's Forest Mensuration....................................8vo, | 4 | 00 |
| Green's Principles of American Forestry.........................12mo, | 1 | 50 |
| Grotenfelt's Principles of Modern Dairy Practice. (Woll.)........12mo, | 2 | 00 |
| * Herrick's Denatured or Industrial Alcohol.......................8vo, | 4 | 00 |
| Kemp and Waugh's Landscape Gardening. (New Edition, Rewritten. In Preparation). | | |
| * McKay and Larsen's Principles and Practice of Butter-making.......8vo, | 1 | 50 |
| Maynard's Landscape Gardening as Applied to Home Decoration......12mo, | 1 | 50 |
| Quaintance and Scott's Insects and Diseases of Fruits. (In Preparation). | | |
| Sanderson's Insects Injurious to Staple Crops.....................12mo, | 1 | 50 |
| * Schwarz's Longleaf Pine in Virgin Forests.......................12mo, | 1 | 25 |
| Stockbridge's Rocks and Soils....................................8vo, | 2 | 50 |
| Winton's Microscopy of Vegetable Foods..........................8vo, | 7 | 50 |
| Woll's Handbook for Farmers and Dairymen......................16mo, | 1 | 50 |

### ARCHITECTURE.

| | | |
|---|---|---|
| Baldwin's Steam Heating for Buildings............................12mo, | 2 | 50 |
| Berg's Buildings and Structures of American Railroads..............4to, | 5 | 00 |
| Birkmire's Architectural Iron and Steel...........................8vo, | 3 | 50 |
| Compound Riveted Girders as Applied in Buildings..............8vo, | 2 | 00 |
| Planning and Construction of American Theatres................8vo, | 3 | 00 |
| Planning and Construction of High Office Buildings.............8vo, | 3 | 50 |
| Skeleton Construction in Buildings............................8vo, | 3 | 00 |
| Briggs's Modern American School Buildings.......................8vo, | 4 | 00 |
| Byrne's Inspection of Material and Wormanship Employed in Construction. 16mo, | 3 | 00 |
| Carpenter's Heating and Ventilating of Buildings..................8vo, | 4 | 00 |

1

| | | |
|---|---|---|
| * Corthell's Allowable Pressure on Deep Foundations............12mo, | 1 | 25 |
| Freitag's Architectural Engineering.................................8vo, | 3 | 50 |
|     Fireproofing of Steel Buildings................................8vo, | 2 | 50 |
| French and Ives's Stereotomy.......................................8vo, | 2 | 50 |
| Gerhard's Guide to Sanitary House-Inspection......................16mo, | 1 | 00 |
| *   Modern Baths and Bath Houses.................................8vo, | 3 | 00 |
|     Sanitation of Public Buildings...............................12mo, | 1 | 50 |
|     Theatre Fires and Panics.....................................12mo, | 1 | 50 |
| Holley and Ladd's Analysis of Mixed Paints, Color Pigments, and Varnishes Large 12mo, | 2 | 50 |
| Johnson's Statics by Algebraic and Graphic Methods................8vo, | 2 | 00 |
| Kellaway's How to Lay Out Suburban Home Grounds..................8vo, | 2 | 00 |
| Kidder's Architects' and Builders' Pocket-book.................16mo, mor., | 5 | 00 |
| Maire's Modern Pigments and their Vehicles.......................12mo, | 2 | 00 |
| Merrill's Non-metallic Minerals: Their Occurrence and Uses.........8vo, | 4 | 00 |
|     Stones for Building and Decoration............................8vo, | 5 | 00 |
| Monckton's Stair-building..........................................4to, | 4 | 00 |
| Patton's Practical Treatise on Foundations.........................8vo, | 5 | 00 |
| Peabody's Naval Architecture.......................................8vo, | 7 | 50 |
| Rice's Concrete-block Manufacture .................................8vo, | 2 | 00 |
| Richey's Handbook for Superintendents of Construction........16mo, mor., | 4 | 00 |
| * Building Mechanics' Ready Reference Book: | | |
|     * Building Foreman's Pocket Book and Ready Reference. (In Preparation). | | |
|     * Carpenters' and Woodworkers' Edition............16mo, mor. | 1 | 50 |
|     * Cement Workers and Plasterer's Edition..........16mo, mor. | 1 | 50 |
|     * Plumbers', Steam-Fitters', and Tinners' Edition.....16mo, mor. | 1 | 50 |
|     * Stone- and Brick-masons' Edition................16mo, mor. | 1 | 50 |
| Sabin's Industrial and Artistic Technology of Paints and Varnish....8vo, | 3 | 00 |
| Siebert and Biggin's Modern Stone-cutting and Masonry..............8vo, | 1 | 50 |
| Snow's Principal Species of Wood...................................8vo, | 3 | 50 |
| Towne's Locks and Builders' Hardware........................18mo, mor. | 3 | 00 |
| Wait's Engineering and Architectural Jurisprudence .................8vo, | 6 | 00 |
|     Sheep, | 6 | 50 |
|     Law of Contracts.............................................8vo, | 3 | 00 |
|     Law of Operations Preliminary to Construction in Engineering and Architecture.................................................8vo, | 5 | 00 |
|     Sheep, | 5 | 50 |
| Wilson's Air Conditioning.........................................12mo, | 1 | 50 |
| Worcester and Atkinson's Small Hospitals, Establishment and Maintenance, Suggestions for Hospital Architecture, with Plans for a Small Hospital. 12mo, | 1 | 25 |

## ARMY AND NAVY.

| | | |
|---|---|---|
| Bernadou's Smokeless Powder, Nitro-cellulose, and the Theory of the Cellulose Molecule....................................................12mo, | 2 | 50 |
| Chase's Art of Pattern Making.....................................12mo, | 2 | 50 |
|     Screw Propellers and Marine Propulsion.........................8vo, | 3 | 00 |
| Cloke's Gunner's Examiner..........................................8vo, | 1 | 50 |
| Craig's Azimuth....................................................4to, | 3 | 50 |
| Crehore and Squier's Polarizing Photo-chronograph..................8vo, | 3 | 00 |
| * Davis's Elements of Law..........................................8vo, | 2 | 50 |
| *   Treatise on the Military Law of United States..................8vo, | 7 | 00 |
|     Sheep, | 7 | 50 |
| De Brack's Cavalry Outpost Duties. (Carr.)..................24mo, mor. | 2 | 00 |
| * Dudley's Military Law and the Procedure of Courts-martial... Large 12mo, | 2 | 50 |
| Durand's Resistance and Propulsion of Ships.......................8vo, | 5 | 00 |

2

| | | |
|---|---|---|
| * Dyer's Handbook of Light Artillery............................12mo, | 3 | 00 |
| Eissler's Modern High Explosives....................................8vo, | 4 | 00 |
| * Fiebeger's Text-book on Field Fortification.................Large 12mo, | 2 | 00 |
| Hamilton and Bond's The Gunner's Catechism.....................18mo, | 1 | 00 |
| * Hoff's Elementary Naval Tactics..................................8vo, | 1 | 50 |
| Ingalls's Handbook of Problems in Direct Fire......................8vo, | 4 | 00 |
| * Lissak's Ordnance and Gunnery....................................8vo, | 6 | 00 |
| * Ludlow's Logarithmic and Trigonometric Tables....................8vo, | 1 | 00 |
| * Lyons's Treatise on Electromagnetic Phenomena. Vols. I. and II..8vo, each, | 6 | 00 |
| * Mahan's Permanent Fortifications. (Mercur.).............8vo, half mor. | 7 | 50 |
| Manual for Courts-martial................................16mo, mor. | 1 | 50 |
| * Mercur's Attack of Fortified Places............................12mo, | 2 | 00 |
| *   Elements of the Art of War...................................8vo, | 4 | 00 |
| Metcalf's Cost of Manufactures—And the Administration of Workshops..8vo, | 5 | 00 |
| *   Ordnance and Gunnery. 2 vols........Text 12mo, Plates atlas form | 5 | 00 |
| Nixon's Adjutants' Manual......................................24mo, | 1 | 00 |
| Peabody's Naval Architecture.......................................8vo, | 7 | 50 |
| * Phelps's Practical Marine Surveying..............................8vo, | 2 | 50 |
| Powell's Army Officer's Examiner................................12mo, | 4 | 00 |
| Sharpe's Art of Subsisting Armies in War.....................18mo, mor. | 1 | 50 |
| * Tupes and Poole's Manual of Bayonet Exercises and Musketry Fencing. 24mo, leather, | | 50 |
| * Weaver's Military Explosives....................................8vo, | 3 | 00 |
| Woodhull's Notes on Military Hygiene..........................16mo, | 1 | 50 |

## ASSAYING.

| | | |
|---|---|---|
| Betts's Lead Refining by Electrolysis..............................8vo, | 4 | 00 |
| Fletcher's Practical Instructions in Quantitative Assaying with the Blowpipe. 16mo, mor. | 1 | 50 |
| Furman's Manual of Practical Assaying............................8vo, | 3 | 00 |
| Lodge's Notes on Assaying and Metallurgical Laboratory Experiments....8vo, | 3 | 00 |
| Low's Technical Methods of Ore Analysis........................ 8vo, | 3 | 00 |
| Miller's Cyanide Process.........................................12mo, | 1 | 00 |
|    Manual of Assaying.........................................12mo, | 1 | 00 |
| Minet's Production of Aluminum and its Industrial Use. (Waldo.).....12mo, | 2 | 50 |
| O'Driscoll's Notes on the Treatment of Gold Ores......................8vo, | 2 | 00 |
| Ricketts and Miller's Notes on Assaying.............................8vo, | 3 | 00 |
| Robine and Lenglen's Cyanide Industry. (Le Clerc.)................8vo, | 4 | 00 |
| Ulke's Modern Electrolytic Copper Refining........................8vo, | 3 | 00 |
| Wilson's Chlorination Process...................................12mo, | 1 | 50 |
|    Cyanide Processes..........................................12mo, | 1 | 50 |

## ASTRONOMY.

| | | |
|---|---|---|
| Comstock's Field Astronomy for Engineers..........................8vo, | 2 | 50 |
| Craig's Azimuth....................................................4to, | 3 | 50 |
| Crandall's Text-book on Geodesy and Least Squares..................8vo, | 3 | 00 |
| Doolittle's Treatise on Practical Astronomy.........................8vo, | 4 | 00 |
| Gore's Elements of Geodesy.......................................8vo, | 2 | 50 |
| Hayford's Text-book of Geodetic Astronomy........................8vo, | 3 | 00 |
| Merriman's Elements of Precise Surveying and Geodesy..............8vo, | 2 | 50 |
| * Michie and Harlow's Practical Astronomy.........................8vo, | 3 | 00 |
| Rust's Ex-meridian Altitude, Azimuth and Star-Finding Tables. (In Press.) | | |
| * White's Elements of Theoretical and Descriptive Astronomy........12mo, | 2 | 00 |

# CHEMISTRY.

| | | |
|---|---|---|
| Abderhalden's Physiological Chemistry in Thirty Lectures. (Hall and Defren). (In Press.) | | |
| * Abegg's Theory of Electrolytic Dissociation. (von Ende.)..........12mo, | 1 | 25 |
| Adriance's Laboratory Calculations and Specific Gravity Tables..:......12mo, | 1 | 25 |
| Alexeyeff's General Principles of Organic Syntheses. (Matthews.)........8vo, | 3 | 00 |
| Allen's Tables for Iron Analysis......................................8vo, | 3 | 00 |
| Arnold's Compendium of Chemistry. (Mandel.)..............Large 12mo, | 3 | 50 |
| Association of State and National Food and Dairy Departments, Hartford Meeting, 1906............................................8vo, | 3 | 00 |
| Jamestown Meeting. 1907.....................................8vo, | 3 | 00 |
| Austen's Notes for Chemical Students ...............................12mo, | 1 | 50 |
| Baskerville's Chemical Elements. (In Preparation). | | |
| Bernadou's Smokeless Powder.—Nitro-cellulose, and Theory of the Cellulose Molecule...............................................12mo, | 2 | 50 |
| * Blanchard's Synthetic Inorganic Chemistry......................12mo, | 1 | 00 |
| * Browning's Introduction to the Rarer Elements....................8vo, | 1 | 50 |
| Brush and Penfield's Manual of Determinative Mineralogy............8vo, | 4 | 00 |
| * Claassen's Beet-sugar Manufacture. (Hall and Rolfe.)......?........8vo, | 3 | 00 |
| Classen's Quantitative Chemical Analysis by Electrolysis. (Boltwood.)..8vo, | 3 | 00 |
| Cohn's Indicators and Test-papers............................12mo, | 2 | 00 |
| Tests and Reagents..........................................8vo, | 3 | 00 |
| * Danneel's Electrochemistry. (Merriam.)........................12mo, | 1 | 25 |
| Duhem's Thermodynamics and Chemistry. (Burgess.)..............8vo, | 4 | 00 |
| Eakle's Mineral Tables for the Determination of Minerals by their Physical Properties.................................................8vo, | 1 | 25 |
| Eissler's Modern High Explosives...................................8vo, | 4 | 00 |
| Effront's Enzymes and their Applications. (Prescott.)...............8vo, | 3 | 00 |
| Erdmann's Introduction to Chemical Preparations. (Dunlap.).......12mo, | 1 | 25 |
| * Fischer's Physiology of Alimentation........................Large 12mo, | 2 | 00 |
| Fletcher's Practical Instructions in Quantitative Assaying with the Blowpipe. 12mo, mor. | 1 | 50 |
| Fowler's Sewage Works Analyses..................................12mo, | 2 | 00 |
| Fresenius's Manual of Qualitative Chemical Analysis. (Wells.).........8vo, | 5 | 00 |
| Manual of Qualitative Chemical Analysis. Part I. Descriptive. (Wells.) 8vo, | 3 | 00 |
| Quantitative Chemical Analysis. (Cohn.) 2 vols................8vo, | 12 | 50 |
| When Sold Separately, Vol. I, $6. Vol. II, $8. | | |
| Fuertes's Water and Public Health................................12mo, | 1 | 50 |
| Furman's Manual of Practical Assaying..............................8vo, | 3 | 00 |
| * Getman's Exercises in Physical Chemistry........................12mo, | 2 | 00 |
| Gill's Gas and Fuel Analysis for Engineers.........................12mo, | 1 | 25 |
| * Gooch and Browning's Outlines of Qualitative Chemical Analysis. Large 12mo, | 1 | 25 |
| Grotenfelt's Principles of Modern Dairy Practice. (Woll.)...........12mo, | 2 | 00 |
| Groth's Introduction to Chemical Crystallography (Marshall).......12mo, | 1 | 25 |
| Hammarsten's Text-book of Physiological Chemistry. (Mandel.).......8vo, | 4 | 00 |
| Hanausek's Microscopy of Technical Products. (Winton.)............8vo, | 5 | 00 |
| * Haskins and Macleod's Organic Chemistry........................12mo, | 2 | 00 |
| Helm's Principles of Mathematical Chemistry. (Morgan.)...........12mo, | 1 | 50 |
| Hering's Ready Reference Tables (Conversion Factors)..........16mo, mor. | 2 | 50 |
| * Herrick's Denatured or Industrial Alcohol.........................8vo, | 4 | 00 |
| Hinds's Inorganic Chemistry.......................................8vo, | 3 | 00 |
| * Laboratory Manual for Students ..............................12mo, | 1 | 00 |
| * Holleman's Laboratory Manual of Organic Chemistry for Beginners. (Walker.)..................................................12mo, | 1 | 00 |
| Text-book of Inorganic Chemistry. (Cooper.)..................8vo, | 2 | 50 |
| Text-book of Organic Chemistry. (Walker and Mott.)..........8vo, | 2 | 50 |
| Holley and Ladd's Analysis of Mixed Paints, Color Pigments, and Varnishes. Large 12mo | 2 | 50 |

| | | |
|---|---|---|
| Hopkins's Oil-chemists' Handbook...................................8vo, | 3 | 00 |
| Iddings's Rock Minerals........................................8vo, | 5 | 00 |
| Jackson's Directions for Laboratory Work in Physiological Chemistry..8vo, | 1 | 25 |
| Johannsen's Determination of Rock-forming Minerals in Thin Sections...8vo, | 4 | 00 |
| Keep's Cast Iron..................................................8vo, | 2 | 50 |
| Ladd's Manual of Quantitative Chemical Analysis...................12mo, | 1 | 00 |
| Landauer's Spectrum Analysis. (Tingle.)..........................8vo, | 3 | 00 |
| * Langworthy and Austen's Occurrence of Aluminium in Vegetable Products, Animal Products, and Natural Waters................8vo, | 2 | 00 |
| Lassar-Cohn's Application of Some General Reactions to Investigations in Organic Chemistry. (Tingle.)..........................12mo, | 1 | 00 |
| Leach's Inspection and Analysis of Food with Special Reference to State Control.........................................................8vo, | 7 | 50 |
| Löb's Electrochemistry of Organic Compounds. (Lorenz.)...........8vo, | 3 | 00 |
| Lodge's Notes on Assaying and Metallurgical Laboratory Experiments....8vo, | 3 | 00 |
| Low's Technical Method of Ore Analysis...........................8vo, | 3 | 00 |
| Lunge's Techno-chemical Analysis. (Cohn.).......................12mo, | 1 | 00 |
| * McKay and Larsen's Principles and Practice of Butter-making.......8vo, | 1 | 50 |
| Maire's Modern Pigments and their Vehicles.......................12mo, | 2 | 00 |
| Mandel's Handbook for Bio-chemical Laboratory....................12mo, | 1 | 50 |
| * Martin's Laboratory Guide to Qualitative Analysis with the Blowpipe..12mo, | | 60 |
| Mason's Examination of Water. (Chemical and Bacteriological.)....12mo, | 1 | 25 |
| Water-supply. (Considered Principally from a Sanitary Standpoint.) 8vo, | 4 | 00 |
| Matthews's The Textile Fibres. 2d Edition, Rewritten..............8vo, | 4 | 00 |
| Meyer's Determination of Radicles in Carbon Compounds. (Tingle.)..12mo, | 1 | 00 |
| Miller's Cyanide Process.........................................12mo, | 1 | 00 |
| Manual of Assaying...........................................12mo, | 1 | 00 |
| Minet's Production of Aluminum and its Industrial Use. (Waldo.)....12mo, | 2 | 50 |
| Mixter's Elementary Text-book of Chemistry.......................12mo, | 1 | 50 |
| Morgan's Elements of Physical Chemistry..........................12mo, | 3 | 00 |
| Outline of the Theory of Solutions and its Results...............12mo, | 1 | 00 |
| * Physical Chemistry for Electrical Engineers...................12mo, | 1 | 50 |
| Morse's Calculations used in Cane-sugar Factories............16mo, mor. | 1 | 50 |
| * Muir's History of Chemical Theories and Laws....................8vo, | 4 | 00 |
| Mulliken's General Method for the Identification of Pure Organic Compounds. Vol. I..........................................Large 8vo, | 5 | 00 |
| O'Driscoll's Notes on the Treatment of Gold Ores..................8vo, | 2 | 00 |
| Ostwald's Conversations on Chemistry. Part One. (Ramsey.).......12mo, | 1 | 50 |
| "       "       "       "       Part Two. (Turnbull.)......12mo, | 2 | 00 |
| * Palmer's Practical Test Book of Chemistry......................12mo, | 1 | 00 |
| * Pauli's Physical Chemistry in the Service of Medicine. (Fischer.)....12mo, | 1 | 25 |
| * Penfield's Notes on Determinative Mineralogy and Record of Mineral Tests. 8vo, paper, | | 50 |
| Tables of Minerals, Including the Use of Minerals and Statistics of Domestic Production.........................................8vo, | 1 | 00 |
| Pictet's Alkaloids and their Chemical Constitution. (Biddle.)........8vo, | 5 | 00 |
| Poole's Calorific Power of Fuels..................................8vo, | 3 | 00 |
| Prescott and Winslow's Elements of Water Bacteriology, with Special Reference to Sanitary Water Analysis.........................12mo, | 1 | 50 |
| * Reisig's Guide to Piece-dyeing...................................8vo, | 25 | 00 |
| Richards and Woodman's Air, Water, and Food from a Sanitary Standpoint..8vo, | 2 | 00 |
| Ricketts and Miller's Notes on Assaying..........................8vo, | 3 | 00 |
| Rideal's Disinfection and the Preservation of Food................8vo, | 4 | 00 |
| Sewage and the Bacterial Purification of Sewage.................8vo, | 4 | 00 |
| Riggs's Elementary Manual for the Chemical Laboratory............8vo, | 1 | 25 |
| Robine and Lenglen's Cyanide Industry. (Le Clerc.)...............8vo, | 4 | 00 |
| Ruddiman's Incompatibilities in Prescriptions.....................8vo, | 2 | 00 |
| Whys in Pharmacy.............................................12mo, | 1 | 00 |

Ruer's Elements of Metallography. (Mathewson). (In Preparation.)
Sabin's Industrial and Artistic Technology of Paints and Varnish........8vo, 3 00
Salkowski's Physiological and Pathological Chemistry. (Orndorff.).....8vo, 2 50
Schimpf's Essentials of Volumetric Analysis. ............................12mo, 1 25
* Qualitative Chemical Analysis..................................8vo, 1 25
    Text-book of Volumetric Analysis. ...............................12mo, 2 50
Smith's Lecture Notes on Chemistry for Dental Students. ...........8vo, 2 50
Spencer's Handbook for Cane Sugar Manufacturers............16mo, mor. 3 00
    Handbook for Chemists of Beet-sugar Houses............ 16mo, mor. 3 00
Stockbridge's Rocks and Soils........................................8vo, 2 50
* Tillman's Descriptive General Chemistry.......................... 8vo, 3 00
*   Elementary Lessons in Heat....................................8vo, 1 50
Treadwell's Qualitative Analysis. (Hall.)............................8vo, 3 00
    Quantitative Analysis. (Hall.)..................................8vo, 4 00
Turneaure and Russell's Public Water-supplies......................8vo, 5 00
Van Deventer's Physical Chemistry for Beginners. (Boltwood.)......12mo, 1 50
Venable's Methods and Devices for Bacterial Treatment of Sewage.......8vo, 3 00
Ward and Whipple's Freshwater Biology. (In Press.)
Ware's Beet-sugar Manufacture and Refining. Vol. I............Small 8vo, 4 00
   "    "    "    "    "     Vol. II. ..........Small 8vo, 5 00
Washington's Manual of the Chemical Analysis of Rocks. .............8vo, 2 00
* Weaver's Military Explosives. ......................................8vo, 3 00
Wells's Laboratory Guide in Qualitative Chemical Analysis............8vo, 1 50
    Short Course in Inorganic Qualitative Chemical Analysis for Engineering
        Students.......................................................12mo, 1 50
    Text-book of Chemical Arithmetic ............................12mo, 1 25
Whipple's Microscopy of Drinking-water..............................8vo, 3 50
Wilson's Chlorination Process........................................12mo, 1 50
    Cyanide Processes ............................................12mo, 1 50
Winton's Microscopy of Vegetable Foods ..........................8vo, 7 50

## CIVIL ENGINEERING.

### BRIDGES AND ROOFS. HYDRAULICS. MATERIALS OF ENGINEERING. RAILWAY ENGINEERING.

Baker's Engineers' Surveying Instruments.........................12mo, 3 00
Bixby's Graphical Computing Table. ..............Paper 19½ × 24¼ inches. 25
Breed and Hosmer's Principles and Practice of Surveying..............8vo, 3 00
* Burr's Ancient and Modern Engineering and the Isthmian Canal ..... 8vo, 3 50
Comstock's Field Astronomy for Engineers........................ ........8vo, 2 50
* Corthell's Allowable Pressures on Deep Foundations..................12mo, 1 25
Crandall's Text-book on Geodesy and Least Squares ..................8vo, 3 00
Davis's Elevation and Stadia Tables.................................8vo, 1 00
Elliott's Engineering for Land Drainage............................12mo, 1 50
    Practical Farm Drainage.........................................12mo, 1 00
* Fiebeger's Treatise on Civil Engineering.............................8vo, 5 00
Flemer's Phototopographic Methods and Instruments.................8vo, 5 00
Folwell's Sewerage. (Designing and Maintenance.)...................8vo, 3 00
Freitag's Architectural Engineering..................................8vo, 3 50
French and Ives's Stereotomy........................................8vo, 2 50
Goodhue's Municipal Improvements.................................12mo, 1 50
Gore's Elements of Geodesy........................................8vo, 2 50
* Hauch and Rice's Tables of Quantities for Preliminary Estimates.......12mo, 1 25
Hayford's Text-book of Geodetic Astronomy........................8vo, 3 00
Hering's Ready Reference Tables (Conversion Factors). .........16mo, mor. 2 50
Howe's Retaining Walls for Earth...................................12mo, 1 25

6

| | | |
|---|---|---|
| * Ives's Adjustments of the Engineer's Transit and Level..........16mo, Bds. | | 25 |
| Ives and Hilts's Problems in Surveying........................16mo, mor. | 1 | 50 |
| Johnson's (J. B.) Theory and Practice of Surveying.............Small 8vo, | 4 | 00 |
| Johnson's (L. J.) Statics by Algebraic and Graphic Methods............8vo, | 2 | 00 |
| Kinnicutt, Winslow and Pratt's Purification of Sewage. (In Preparation). | | |
| Laplace's Philosophical Essay on Probabilities. (Truscott and Emory.) | | |
| 12mo, | 2 | 00 |
| Mahan's Descriptive Geometry........................................8vo, | 1 | 50 |
| Treatise on Civil Engineering. (1873.) (Wood.)...............8vo, | 5 | 00 |
| Merriman's Elements of Precise Surveying and Geodesy...............8vo, | 2 | 50 |
| Merriman and Brooks's Handbook for Surveyors.............16mo, mor. | 2 | 00 |
| Morrison's Elements of Highway Engineering. (In Press.) | | |
| Nugent's Plane Surveying...........................................8vo, | 3 | 50 |
| Ogden's Sewer Design..............................................12mo, | 2 | 00 |
| Parsons's Disposal of Municipal Refuse..............................8vo, | 2 | 00 |
| Patton's Treatise on Civil Engineering..................8vo, half leather, | 7 | 50 |
| Reed's Topographical Drawing and Sketching.........................4to, | 5 | 00 |
| Rideal's Sewage and the Bacterial Purification of Sewage............8vo, | 4 | 00 |
| Riemer's Shaft-sinking under Difficult Conditions. (Corning and Peele.)..8vo, | 3 | 00 |
| Siebert and Biggin's Modern Stone-cutting and Masonry...............8vo, | 1 | 50 |
| Smith's Manual of Topographical Drawing. (McMillan.).............8vo, | 2 | 50 |
| Soper's Air and Ventilation of Subways. (In Press.) | | |
| Tracy's Plane Surveying........................................16mo, mor. | 3 | 00 |
| * Trautwine's Civil Engineer's Pocket-book....................16mo, mor. | 5 | 00 |
| Venable's Garbage Crematories in America..........................8vo, | 2 | 00 |
| Methods and Devices for Bacterial Treatment of Sewage........8vo, | 3 | 00 |
| Wait's Engineering and Architectural Jurisprudence.................8vo, | 6 | 00 |
| Sheep, | 6 | 50 |
| Law of Contracts..............................................8vo, | 3 | 00 |
| Law of Operations Preliminary to Construction in Engineering and Architecture...................................................8vo, | 5 | 00 |
| Sheep, | 5 | 50 |
| Warren's Stereotomy—Problems in Stone-cutting....................8vo, | 2 | 50 |
| * Waterbury's Vest-Pocket Hand-book of Mathematics for Engineers. | | |
| 2⅞ × 5⅜ inches, mor. | 1 | 00 |
| Webb's Problems in the Use and Adjustment of Engineering Instruments. | | |
| 16mo, mor. | 1 | 25 |
| Wilson's Topographic Surveying....................................8vo, | 3 | 50 |

## BRIDGES AND ROOFS.

| | | |
|---|---|---|
| Boller's Practical Treatise on the Construction of Iron Highway Bridges..8vo, | 2 | 00 |
| Burr and Falk's Design and Construction of Metallic Bridges...........8vo, | 5 | 00 |
| Influence Lines for Bridge and Roof Computations................8vo, | 3 | 00 |
| Du Bois's Mechanics of Engineering. Vol. II....................Small 4to, | 10 | 00 |
| Foster's Treatise on Wooden Trestle Bridges..........................4to, | 5 | 00 |
| Fowler's Ordinary Foundations.....................................8vo, | 3 | 50 |
| French and Ives's Stereotomy......................................8vo, | 2 | 50 |
| Greene's Arches in Wood, Iron, and Stone..........................8vo, | 2 | 50 |
| Bridge Trusses..............................................8vo, | 2 | 50 |
| Roof Trusses................................................8vo, | 1 | 25 |
| Grimm's Secondary Stresses in Bridge Trusses......................8vo, | 2 | 50 |
| Heller's Stresses in Structures and the Accompanying Deformations......8vo, | | |
| Howe's Design of Simple Roof-trusses in Wood and Steel.............8vo, | 2 | 00 |
| Symmetrical Masonry Arches..................................8vo, | 2 | 50 |
| Treatise on Arches..........................................8vo, | 4 | 00 |
| Johnson, Bryan, and Turneaure's Theory and Practice in the Designing of Modern Framed Structures........................Small 4to, | 10 | 00 |

Merriman and Jacoby's Text-book on Roofs and Bridges:
    Part I.   Stresses in Simple Trusses..............................8vo, 2 50
    Part II.  Graphic Statics......................................8vo, 2 50
    Part III. Bridge Design.......................................8vo, 2 50
    Part IV.  Higher Structures...................................8vo, 2 50
Morison's Memphis Bridge................................Oblong 4to, 10 00
Sondericker's Graphic Statics, with Applications to Trusses, Beams, and Arches.
                                                                        8vo, 2 00
Waddell's De Pontibus, Pocket-book for Bridge Engineers...... 16mo, mor, 2 00
*    Specifications for Steel Bridges...............................12mo, 50
Waddell and Harrington's Bridge Engineering. (In Preparation.)
Wright's Designing of Draw-spans. Two parts in one volume..........8vo, 3 50


# HYDRAULICS.

Barnes's Ice Formation..............................................8vo, 3 00
Bazin's Experiments upon the Contraction of the Liquid Vein Issuing from
    an Orifice. (Trautwine.)....................................8vo, 2 00
Bovey's Treatise on Hydraulics......................................8vo, 5 00
Church's Diagrams of Mean Velocity of Water in Open Channels.
                                                               Oblong 4to, paper, 1 50
    Hydraulic Motors...........................................8vo, 2 00
    Mechanics of Engineering....................................8vo, 6 00
Coffin's Graphical Solution of Hydraulic Problems.........16mo, morocco, 2 50
Flather's Dynamometers, and the Measurement of Power............12mo, 3 00
Folwell's Water-supply Engineering.................................8vo, 4 00
Frizell's Water-power..............................................8vo, 5 00
Fuertes's Water and Public Health................................12mo, 1 50
    Water-filtration Works.....................................12mo, 2 50
Ganguillet and Kutter's General Formula for the Uniform Flow of Water in
    Rivers and Other Channels. (Hering and Trautwine.).........8vo, 4 00
Hazen's Clean Water and How to Get It..................... Large 12mo, 1 50
    Filtration of Public Water-supplies.........................8vo, 3 00
Hazlehurst's Towers and Tanks for Water-works.....................8vo, 2 50
Herschel's 115 Experiments on the Carrying Capacity of Large, Riveted, Metal
    Conduits..................................................8vo, 2 00
Hoyt and Grover's River Discharge..................................8vo, 2 00
Hubbard and Kiersted's Water-works Management and Maintenance.....8vo, 4 00
* Lyndon's Development and Electrical Distribution of Water Power....8vo, 3 00
Mason's Water-supply. (Considered Principally from a Sanitary Standpoint.)
                                                                                 8vo, 4 00
Merriman's Treatise on Hydraulics..................................8vo, 5 00
* Michie's Elements of Analytical Mechanics........................8vo, 4 00
Molitor's Hydraulics of Rivers, Weirs and Sluices. (In Press.)
Schuyler's Reservoirs for Irrigation, Water-power, and Domestic Water-
    supply...............................................Large 8vo, 5 00
* Thomas and Watt's Improvement of Rivers.........................4to, 6 00
Turneaure and Russell's Public Water-supplies.....................8vo, 5 00
Wegmann's Design and Construction of Dams. 5th Ed., enlarged.....4to, 6 00
    Water-supply of the City of New York from 1658 to 1895..........4to, 10 00
Whipple's Value of Pure Water............................Large 12mo, 1 00
Williams and Hazen's Hydraulic Tables.............................8vo, 1 50
Wilson's Irrigation Engineering.................................Small 8vo, 4 00
Wolff's Windmill as a Prime Mover.................................8vo, 3 00
Wood's Elements of Analytical Mechanics...........................8vo, 3 00
    Turbines...................................................8vo, 2 50

## MATERIALS OF ENGINEERING.

| | | |
|---|---|---|
| Baker's Roads and Pavements...........................................8vo, | 5 | 00 |
|     Treatise on Masonry Construction.................................8vo, | 5 | 00 |
| Birkmire's Architectural Iron and Steel..................................8vo, | 3 | 50 |
|     Compound Riveted Girders as Applied in Buildings............8vo, | 2 | 00 |
| Black's United States Public Works .........................Oblong 4to, | 5 | 00 |
| Bleininger's Manufacture of Hydraulic Cement. (In Preparation.) | | |
| * Bovey's Strength of Materials and Theory of Structures..............8vo, | 7 | 50 |
| Burr's Elasticity and Resistance of the Materials of Engineering........8vo, | 7 | 50 |
| Byrne's Highway Construction........................................8vo, | 5 | 00 |
|     Inspection of the Materials and Workmanship Employed in Construction. | | |
|     16mo, | 3 | 00 |
| Church's Mechanics of Engineering....................................8vo, | 6 | 00 |
| Du Bois's Mechanics of Engineering. | | |
|     Vol. I. Kinematics, Statics, Kinetics......................Small 4to, | 7 | 50 |
|     Vol. II. The Stresses in Framed Structures, Strength of Materials and | | |
|         Theory of Flexures.........................................Small 4to, | 10 | 00 |
| *Eckel's Cements, Limes, and Plasters..................................8vo, | 6 | 00 |
|     Stone and Clay Products used in Engineering. (In Preparation.) | | |
| Fowler's Ordinary Foundations........................................8vo, | 3 | 50 |
| Graves's Forest Mensuration..........................................8vo, | 4 | 00 |
| Green's Principles of American Forestry..............................12mo, | 1 | 50 |
| * Greene's Structural Mechanics.......................................8vo, | 2 | 50 |
| Holly and Ladd's Analysis of Mixed Paints, Color Pigments and Varnishes | | |
|     Large 12mo, | 2 | 50 |
| Johnson's Materials of Construction..........................Large 8vo, | 6 | 00 |
| Keep's Cast Iron....................................................8vo, | 2 | 50 |
| Kidder's Architects and Builders' Pocket-book.......................16mo, | 5 | 00 |
| Lanza's Applied Mechanics............................................8vo, | 7 | 50 |
| Maire's Modern Pigments and their Vehicles .......................12mo, | 2 | 00 |
| Martens's Handbook on Testing Materials. (Henning.) 2 vols........8vo, | 7 | 50 |
| Maurer's Technical Mechanics.........................................8vo, | 4 | 00 |
| Merrill's Stones for Building and Decoration.......................... 8vo, | 5 | 00 |
| Merriman's Mechanics of Materials....................................8vo, | 5 | 00 |
| *    Strength of Materials ........................................12mo, | 1 | 00 |
| Metcalf's Steel. A Manual for Steel-users..........................12mo, | 2 | 00 |
| Patton's Practical Treatise on Foundations............................8vo, | 5 | 00 |
| Rice's Concrete Block Manufacture....................................8vo, | 2 | 00 |
| Richardson's Modern Asphalt Pavements..............................8vo, | 3 | 00 |
| Richey's Handbook for Superintendents of Co tr tion......16mo, mor., | 4 | 00 |
| * Ries's Clays: Their Occurrence, Properties, and Uses...............8vo, | 5 | 00 |
| Sabin's Industrial and Artistic Technology of Paints and Varnish........8vo, | 3 | 00 |
| * Schwarz's Longleaf Pine in Virgin Forest..........................12mo, | 1 | 25 |
| Snow's Principal Species of Wood....................................8vo, | 3 | 50 |
| Spalding's Hydraulic Cement.........................................12mo, | 2 | 00 |
|     Text-book on Roads and Pavements............................12mo, | 2 | 00 |
| Taylor and Thompson's Treatise on Concrete, Plain and Reinforced......8vo, | 5 | 00 |
| Thurston's Materials of Engineering. In Three Parts..................8vo, | 8 | 00 |
|     Part I. Non-metallic Materials of Engineering and Metallurgy.....8vo, | 2 | 00 |
|     Part II. Iron and Steel........................................8vo, | 3 | 50 |
|     Part III. A Treatise on Brasses, Bronzes, and Other Alloys and their | | |
|         Constituents.................................................8vo, | 2 | 50 |
| Tillson's Street Pavements and Paving Materials......................8vo, | 4 | 00 |
| Turneaure and Maurer's Principles of Reinforced Concrete Construction...8vo, | 3 | 00 |
| Wood's (De V.) Treatise on the Resistance of Materials, and an Appendix on | | |
|     the Preservation of Timber.......................................8vo, | 2 | 00 |
| Wood's (M. P.) Rustless Coatings: Corrosion and Electrolysis of Iron and | | |
|     Steel..........................................................8vo, | 4 | 00 |

## RAILWAY ENGINEERING.

| | | |
|---|---|---|
| Andrews's Handbook for Street Railway Engineers........3x5 inches, mor. | 1 | 25 |
| Berg's Buildings and Structures of American Railroads................4to, | 5 | 00 |
| Brooks's Handbook of Street Railroad Location...............16mo, mor. | 1 | 50 |
| Butt's Civil Engineer's Field-book............................16mo, mor. | 2 | 50 |
| Crandall's Railway and Other Earthwork Tables....................8vo, | 1 | 50 |
|     Transition Curve.........................................16mo, mor. | 1 | 50 |
| *Crockett's Methods for Earthwork Computations....................8vo, | 1 | 50 |
| Dawson's "Engineering" and Electric Traction Pocket-book......16mo, mor. | 5 | 00 |
| Dredge's History of the Pennsylvania Railroad: (1879)..............Paper, | 5 | 00 |
| Fisher's Table of Cubic Yards.................................Cardboard, | | 25 |
| Godwin's Railroad Engineers' Field-book and Explorers' Guide...16mo, mor. | 2 | 50 |
| Hudson's Tables for Calculating the Cubic Contents of Excavations and Embankments..................................................8vo, | 1 | 00 |
| Ives and Hilts's Problems in Surveying, Railroad Surveying and Geodesy  16mo, mor. | 1 | 50 |
| Molitor and Beard's Manual for Resident Engineers................16mo, | 1 | 00 |
| Nagle's Field Manual for Railroad Engineers..................16mo, mor. | 3 | 00 |
| Philbrick's Field Manual for Engineers........................16mo, mor. | 3 | 00 |
| Raymond's Railroad Engineering. 3 volumes. | | |
|     Vol. I. Railroad Field Geometry. (In Preparation.) | | |
|     Vol. II. Elements of Railroad Engineering....................8vo, | 3 | 50 |
|     Vol. III. Railroad Engineer's Field Book. (In Preparation.) | | |
| Searles's Field Engineering..................................16mo, mor. | 3 | 00 |
|     Railroad Spiral...........................................16mo, mor. | 1 | 50 |
| Taylor's Prismoidal Formulæ and Earthwork........................8vo, | 1 | 50 |
| *Trautwine's Field Practice of Laying Out Circular Curves for Railroads.  12mo. mor, | 2 | 50 |
| *   Method of Calculating the Cubic Contents of Excavations and Embankments by the Aid of Diagrams................................8vo, | 2 | 00 |
| Webb's Economics of Railroad Construction.................Large 12mo, | 2 | 50 |
|     Railroad Construction....................................16mo, mor. | 5 | 00 |
| Wellington's Economic Theory of the Location of Railways.......Small 8vo, | 5 | 00 |

## DRAWING.

| | | |
|---|---|---|
| Barr's Kinematics of Machinery....................................8vo, | 2 | 50 |
| *Bartlett's Mechanical Drawing....................................8vo, | 3 | 00 |
| *   "   "   "   Abridged Ed......................8vo, | 1 | 50 |
| Coolidge's Manual of Drawing............................8vo, paper, | 1 | 00 |
| Coolidge and Freeman's Elements of General Drafting for Mechanical Engineers............................................Oblong 4to, | 2 | 50 |
| Durley's Kinematics of Machines..................................8vo, | 4 | 00 |
| Emch's Introduction to Projective Geometry and its Applications........8vo, | 2 | 50 |
| Hill's Text-book on Shades and Shadows, and Perspective.............8vo, | 2 | 00 |
| Jamison's Advanced Mechanical Drawing..........................8vo, | 2 | 00 |
|     Elements of Mechanical Drawing..............................8vo, | 2 | 50 |
| Jones's Machine Design: | | |
|     Part I. Kinematics of Machinery............................8vo, | 1 | 50 |
|     Part II. Form, Strength, and Proportions of Parts...............8vo, | 3 | 00 |
| MacCord's Elements of Descriptive Geometry......................8vo, | 3 | 00 |
|     Kinematics; or, Practical Mechanism........................8vo, | 5 | 00 |
|     Mechanical Drawing.......................................4to, | 4 | 00 |
|     Velocity Diagrams.........................................8vo, | 1 | 50 |
| McLeod's Descriptive Geometry...........................Large 12mo, | 1 | 50 |
| *Mahan's Descriptive Geometry and Stone-cutting...................8vo, | 1 | 50 |
|     Industrial Drawing. (Thompson.)............................8vo, | 3 | 50 |

| | | |
|---|---|---|
| Moyer's Descriptive Geometry.........................8vo, | 2 | 00 |
| Reed's Topographical Drawing and Sketching....................4to, | 5 | 00 |
| Reid's Course in Mechanical Drawing.........................8vo, | 2 | 00 |
|     Text-book of Mechanical Drawing and Elementary Machine Design.8vo, | 3 | 00 |
| Robinson's Principles of Mechanism............................8vo, | 3 | 00 |
| Schwamb and Merrill's Elements of Mechanism...................8vo, | 3 | 00 |
| Smith's (R. S.) Manual of Topographical Drawing. (McMillan.).......8vo, | 2 | 50 |
| Smith (A. W.) and Marx's Machine Design........................8vo, | 3 | 00 |
| * Titsworth's Elements of Mechanical Drawing..............Oblong 8vo, | 1 | 25 |
| Warren's Drafting Instruments and Operations.....................12mo, | 1 | 25 |
|     Elements of Descriptive Geometry, Shadows, and Perspective.......8vo, | 3 | 50 |
|     Elements of Machine Construction and Drawing.................8vo, | 7 | 50 |
|     Elements of Plane and Solid Free-hand Geometrical Drawing....12mo, | 1 | 00 |
|     General Problems of Shades and Shadows.....................8vo, | 3 | 00 |
|     Manual of Elementary Problems in the Linear Perspective of Form and Shadow..................................................12mo, | 1 | 00 |
|     Manual of Elementary Projection Drawing.....................12mo, | 1 | 50 |
|     Plane Problems in Elementary Geometry......................12mo, | 1 | 25 |
|     Problems, Theorems, and Examples in Descriptive Geometry.......8vo, | 2 | 50 |
| Weisbach's Kinematics and Power of Transmission. (Hermann and Klein.)...................................................8vo, | 5 | 00 |
| Wilson's (H. M.) Topographic Surveying.........................8vo, | 3 | 50 |
| Wilson's (V. T.) Free-hand Lettering...........................8vo, | 1 | 00 |
|     Free-hand Perspective....................................8vo, | 2 | 50 |
| Woolf's Elementary Course in Descriptive Geometry............Large 8vo, | 3 | 00 |

## ELECTRICITY AND PHYSICS.

| | | |
|---|---|---|
| * Abegg's Theory of Electrolytic Dissociation. (von Ende.).........12mo, | 1 | 25 |
| Andrews's Hand-Book for Street Railway Engineering.....3×5 inches, mor., | 1 | 25 |
| Anthony and Brackett's Text-book of Physics. (Magie.).......Large 12mo, | 3 | 00 |
| Anthony's Lecture-notes on the Theory of Electrical Measurements....12mo, | 1 | 00 |
| Benjamin's History of Electricity............................8vo, | 3 | 00 |
|     Voltaic Cell..........................................8vo, | 3 | 00 |
| Betts's Lead Refining and Electrolysis...........................8vo, | 4 | 00 |
| Classen's Quantitative Chemical Analysis by Electrolysis. (Boltwood.).8vo, | 3 | 00 |
| * Collins's Manual of Wireless Telegraphy.......................12mo, | 1 | 50 |
|     Mor. | 2 | 00 |
| Crehore and Squier's Polarizing Photo-chronograph...................8vo, | 3 | 00 |
| * Danneel's Electrochemistry. (Merriam.)......................12mo, | 1 | 25 |
| Dawson's "Engineering" and Electric Traction Pocket-book.....16mo, mor | 5 | 00 |
| Dolezalek's Theory of the Lead Accumulator (Storage Battery). (von Ende.) 12mo, | 2 | 50 |
| Duhem's Thermodynamics and Chemistry. (Burgess.)...............8vo, | 4 | 00 |
| Flather's Dynamometers, and the Measurement of Power...........12mo, | 3 | 00 |
| Gilbert's De Magnete. (Mottelay.)...........................8vo, | 2 | 50 |
| * Hanchett's Alternating Currents............................12mo, | 1 | 00 |
| Hering's Ready Reference Tables (Conversion Factors)..........16mo, mor. | 2 | 50 |
| Hobart and Ellis's High-speed Dynamo Electric Machinery. (In Press.) | | |
| Holman's Precision of Measurements............................8vo, | 2 | 00 |
|     Telescopic Mirror-scale Method, Adjustments, and Tests....Large 8vo, | | 75 |
| * Karapetoff's Experimental Electrical Engineering..................8vo, | 6 | 00 |
| Kinzbrunner's Testing of Continuous-current Machines..............8vo, | 2 | 00 |
| Landauer's Spectrum Analysis. (Tingle.).......................8vo, | 3 | 00 |
| Le Chatelier's High-temperature Measurements. (Boudouard—Burgess.) 12mo, | 3 | 00 |
| Löb's Electrochemistry of Organic Compounds. (Lorenz.)...........8vo, | 3 | 00 |
| * Lyndon's Development and Electrical Distribntion of Water Power....8vo, | 3 | 00 |
| * Lyons's Treatise on Electromagnetic Phenomena. Vols. I. and II. 8vo, each, | 6 | 00 |
| * Michie's Elements of Wave Motion Relating to Sound and Light......8vo, | 4 | 00 |

Morgan's Outline of the Theory of Solution and its Results............12mo, 1 00
*    Physical Chemistry for Electrical Engineers...................12mo, 1 50
Niaudet's Elementary Treatise on Electric Batteries. (Fishback)....12mo, 2 50
* Norris's Introduction to the Study of Electrical Engineering..........8vo, 2 50
* Parshall and Hobart's Electric Machine Design.........4to, half morocco, 12 50
Reagan's Locomotives: Simple, Compound, and Electric. New Edition.
                                                                             Large 12mo, 3 50
* Rosenberg's Electrical Engineering. (Haldane Gee—Kinzbrunner.)...8vo, 2 00
Ryan, Norris, and Hoxie's Electrical Machinery. Vol. I..............8vo, 2 50
Schapper's Laboratory Guide for Students in Physical Chemistry......12mo, 1 00
Thurston's Stationary Steam-engines................................8vo, 2 50
* Tillman's Elementary Lessons in Heat..............................8vo, 1 50
Tory and Pitcher's Manual of Laboratory Physics...........Large 12mo, 2 00
Ulke's Modern Electrolytic Copper Refining.........................8vo, 3 00

## LAW.

* Davis's Elements of Law........................................8vo, 2 50
*    Treatise on the Military Law of United States...................8vo, 7 00
*                                                                                                             Sheep, 7 50
* Dudley's Military Law and the Procedure of Courts-martial....Large 12mo, 2 50
Manual for Courts-martial...................................16mo, mor. 1 50
Wait's Engineering and Architectural Jurisprudence.................8vo, 6 00
                                                                                                                  Sheep, 6 50
   Law of Contracts...........................................8vo, 3 00
   Law of Operations Preliminary to Construction in Engineering and Architecture....................................................8vo, 5 00
                                                                                                                   Sheep, 5 50

## MATHEMATICS.

Baker's Elliptic Functions........................................8vo, 1 50
Briggs's Elements of Plane Analytic Geometry. (Bôcher)............12mo, 1 00
* Buchanan's Plane and Spherical Trigonometry.....................8vo, 1 00
Byerley's Harmonic Functions....................................8vo, 1 00
Chandler's Elements of the Infinitesimal Calculus..................12mo, 2 00
Compton's Manual of Logarithmic Computations...................12mo, 1 50
Davis's Introduction to the Logic of Algebra.......................8vo, 1 50
* Dickson's College Algebra.................................Large 12mo, 1 50
*    Introduction to the Theory of Algebraic Equations........Large 12mo, 1 25
Emch's Introduction to Projective Geometry and its Applications......8vo, 2 50
Fiske's Functions of a Complex Variable............................8vo, 1 00
Halsted's Elementary Synthetic Geometry..........................8vo, 1 50
   Elements of Geometry......................................8vo, 1 75
*    Rational Geometry.........................................12mo, 1 50
Hyde's Grassmann's Space Analysis...............................8vo, 1 00
* Jonnson's (J B.) Three-place Logarithmic Tables: Vest-pocket size, paper, 15
                                                                                       100 copies, 5 00
*                  Mounted on heavy cardboard, 8 × 10 inches, 25
                                                                                        10 copies, 2 00
Johnson's (W. W.) Abridged Editions of Differential and Integral Calculus
                                                                         Large 12mo, 1 vol. 2 50
   Curve Tracing in Cartesian Co-ordinates....................12mo, 1 00
   Differential Equations......................................8vo, 1 00
   Elementary Treatise on Differential Calculus. (In Press.)
   Elementary Treatise on the Integral Calculus...........Large 12mo, 1 50
*    Theoretical Mechanics......................................12mo, 3 00
   Theory of Errors and the Method of Least Squares............12mo, 1 50
   Treatise on Differential Calculus......................Large 12mo, 3 00
   Treatise on the Integral Calculus.......................Large 12mo, 3 00
   Treatise on Ordinary and Partial Differential Equations..Large 12mo, 3 50

12

| | | |
|---|---|---|
| Laplace's Philosophical Essay on Probabilities. (Truscott and Emory.).12mo, | 2 | 00 |
| * Ludlow and Bass's Elements of Trigonometry and Logarithmic and Other Tables ................8vo, | 3 | 00 |
| Trigonometry and Tables published separately ..................Each, | 2 | 00 |
| * Ludlow's Logarithmic and Trigonometric Tables ....................8vo, | 1 | 00 |
| Macfarlane's Vector Analysis and Quaternions......................8vo, | 1 | 00 |
| McMahon's Hyperbolic Functions................................8vo, | 1 | 00 |
| Manning's Irrational Numbers and their Representation by Sequences and Series 12mo, | 1 | 25 |
| Mathematical Monographs. Edited by Mansfield Merriman and Robert S. Woodward......................................Octavo, each | 1 | 00 |

No. 1. History of Modern Mathematics, by David Eugene Smith. No. 2. Synthetic Projective Geometry, by George Bruce Halsted. No. 3. Determinants. by Laenas Gifford Weld. No. 4. Hyperbolic Functions, by James McMahon. No. 5. Harmonic Functions, by William E. Byerly. No. 6. Grassmann's Space Analysis, by Edward W. Hyde. No. 7. Probability and Theory of Errors, by Robert S. Woodward. No. 8. Vector Analysis and Quaternions, by Alexander. Macfarlane. No. 9. Differential Equations, by William Woolsey Johnson. No. 10. The Solution of Equations, by Mansfield Merriman. No. 11. Functions of a Complex Variable, by Thomas S. Fiske.

| | | |
|---|---|---|
| Maurer's Technical Mechanics.......................................8vo, | 4 | 00 |
| Merriman's Method of Least Squares...........................8vo, | 2 | 00 |
| Solution of Equations .........................................8vo, | 1 | 00 |
| Rice and Johnson's Differential and Integral Calculus. 2 vols. in one. Large 12mo, | 1 | 50 |
| Elementary Treatise on the Differential Calculus...........Large 12mo, | 3 | 00 |
| Smith's History of Modern Mathematics ..........................8vo, | 1 | 00 |
| * Veblen and Lennes's Introduction to the Real Infinitesimal Analysis of One Variable ...............................................8vo, | 2 | 00 |
| * Waterbury's Vest Pocket Hand-Book of Mathematics for Engineers. 2⅞ × 5¾ inches, mor., | 1 | 00 |
| Weld's Determinations...........................................8vo, | 1 | 00 |
| Wood's Elements of Co-ordinate Geometry......................8vo, | 2 | 00 |
| Woodward's Probability and Theory of Errors......................8vo, | 1 | 00 |

## MECHANICAL ENGINEERING.

### MATERIALS OF ENGINEERING, STEAM-ENGINES AND BOILERS.

| | | |
|---|---|---|
| Bacon's Forge Practice...........................................12mo, | 1 | 50 |
| Baldwin's Steam Heating for Buildings.............................12mo, | 2 | 50 |
| Barr's Kinematics of Machinery....................................8vo, | 2 | 50 |
| * Bartlett's Mechanical Drawing.....................................8vo, | 3 | 00 |
| *  "       "       "    Abridged Ed........................8vo, | 1 | 50 |
| Benjamin's Wrinkles and Recipes...................................12mo, | 2 | 00 |
| * Burr's Ancient and Modern Engineering and the Isthmian Canal......8vo, | 3 | 50 |
| Carpenter's Experimental Engineering.............................8vo, | 6 | 00 |
| Heating and Ventilating Buildings..............................8vo, | 4 | 00 |
| Clerk's Gas and Oil Engine.................................. Large 12mo, | 4 | 00 |
| Compton's First Lessons in Metal Working ........................12mo, | 1 | 50 |
| Compton and De Groodt's Speed Lathe............................12mo, | 1 | 50 |
| Coolidge's Manual of Drawing................................8vo, paper, | 1 | 00 |
| Coolidge and Freeman's Elements of General Drafting for Mechanical Engineers.....................................Oblong 4to, | 2 | 50 |
| Cromwell's Treatise on Belts and Pulleys...........................12mo, | 1 | 50 |
| Treatise on Toothed Gearing....................................12mo, | 1 | 50 |
| Durley's Kinematics of Machines..................................8vo, | 4 | 00 |

| | | |
|---|---|---|
| Flather's Dynamometers and the Measurement of Power............12mo, | 3 | 00 |
| Rope Driving............................................................12mo, | 2 | 00 |
| Gill's Gas and Fuel Analysis for Engineers........................12mo, | 1 | 25 |
| Goss's Locomotive Sparks.................................................8vo, | 2 | 00 |
| Hall's Car Lubrication..................................................12mo, | 1 | 00 |
| Hering's Ready Reference Tables (Conversion Factors)..........16mo, mor., | 2 | 50 |
| Hobart and Ellis's High Speed Dynamo Electric Machinery. (In Press.) | | |
| Hutton's Gas Engine.......................................................8vo, | 5 | 00 |
| Jamison's Advanced Mechanical Drawing................................8vo, | 2 | 00 |
| Elements of Mechanical Drawing............................8vo, | 2 | 50 |
| Jones's Machine Design: | | |
| Part I. Kinematics of Machinery....................................8vo, | 1 | 50 |
| Part II. Form, Strength, and Proportions of Parts................8vo, | 3 | 00 |
| Kent's Mechanical Engineers' Pocket-book....................16mo, mor, | 5 | 00 |
| Kerr's Power and Power Transmission...................................8vo, | 2 | 00 |
| Leonard's Machine Shop Tools and Methods...........................8vo, | 4 | 00 |
| * Lorenz's Modern Refrigerating Machinery. (Pope, Haven, and Dean.)..8vo, | 4 | 00 |
| MacCord's Kinematics; or, Practical Mechanism.......................8vo, | 5 | 00 |
| Mechanical Drawing....................................................4to, | 4 | 00 |
| Velocity Diagrams....................................................8vo, | 1 | 50 |
| MacFarland's Standard Reduction Factors for Gases....................8vo, | 1 | 50 |
| Mahan's Industrial Drawing. *(Thompson.)............................8vo, | 3 | 50 |
| * Parshall and Hobart's Electric Machine Design.....Small 4to, half leather, | 12 | 50 |
| Peele's Compressed Air Plant for Mines. (In Press.) | | |
| Poole's Calorific Power of Fuels.........................................8vo, | 3 | 00 |
| * Porter's Engineering Reminiscences, 1855 to 1882...................8vo, | 3 | 00 |
| Reid's Course in Mechanical Drawing....................................8vo, | 2 | 00 |
| Text-book of Mechanical Drawing and Elementary Machine Design.8vo, | 3 | 00 |
| Richard's Compressed Air .............................................12mo, | 1 | 50 |
| Robinson's Principles of Mechanism......................................8vo, | 3 | 00 |
| Schwamb and Merrill's Elements of Mechanism.......................8vo, | 3 | 00 |
| Smith's (O.) Press-working of Metals....................................8vo, | 3 | 00 |
| Smith (A. W.) and Marx's Machine Design............................8vo, | 3 | 00 |
| Thurston's Animal as a Machine and Prime Motor, and the Laws of Energetics. | | |
| 12mo, | 1 | 00 |
| Treatise on Friction and Lost Work in Machinery and Mill Work...8vo, | 3 | 00 |
| Tillson's Complete Automobile Instructor ..........................16mo, | 1 | 50 |
| mor., | 2 | 00 |
| * Titsworth's Elements of Mechanical Drawing.................Oblong 8vo, | 1 | 25 |
| Warren's Elements of Machine Construction and Drawing............8vo, | 7 | 50 |
| * Waterbury's Vest Pocket Hand Book of Mathematics for Engineers. | | |
| $2\frac{7}{8} \times 5\frac{3}{8}$ inches, mor., | 1 | 00 |
| Weisbach's Kinematics and the Power of Transmission. (Herrmann—Klein.).............................................................8vo, | 5 | 00 |
| Machinery of Transmission and Governors. (Herrmann—Klein.)..8vo, | 5 | 00 |
| Wolff's Windmill as a Prime Mover.....................................8vo, | 3 | 00 |
| Wood's Turbines..........................................................8vo, | 2 | 50 |

## MATERIALS OF ENGINEERING.

| | | |
|---|---|---|
| * Bovey's Strength of Materials and Theory of Structures.............8vo, | 7 | 50 |
| Burr's Elasticity and Resistance of the Materials of Engineering........8vo, | 7 | 50 |
| Church's Mechanics of Engineering....................................8vo, | 6 | 00 |
| * Greene's Structural Mechanics.......................................8vo, | 2 | 50 |
| Holley and Ladd's Analysis of Mixed Paints, Color Pigments, and Varnishes. | | |
| Large 12mo, | 2 | 50 |
| Johnson's Materials of Construction.................................. 8vo, | 6 | 00 |
| Keep's Cast Iron.........................................................8vo, | 2 | 50 |
| Lanza's Applied Mechanics..............................................8vo, | 7 | 50 |

| | | |
|---|---|---|
| Maire's Modern Pigments and their Vehicles......................12mo, | 2 | 00 |
| Martens's Handbook on Testing Materials. (Henning.)...............8vo, | 7 | 50 |
| Maurer's Technical Mechanics.....................................8vo, | 4 | 00 |
| Merriman's Mechanics of Materials................................8vo, | 5 | 00 |
|    *    Strength of Materials.........................................12mo, | 1 | 00 |
| Metcalf's Steel. A Manual for Steel-users............................12mo, | 2 | 00 |
| Sabin's Industrial and Artistic Technology of Paints and Varnish........8vo, | 3 | 00 |
| Smith's Materials of Machines....................................12mo, | 1 | 00 |
| Thurston's Materials of Engineering.........................3 vols., 8vo, | 8 | 00 |
|    Part I. Non-metallic Materials of Engineering, see Civil Engineering, page 9. | | |
|    Part II. Iron and Steel..........................................8vo, | 3 | 50 |
|    Part III. A Treatise on Brasses, Bronzes, and Other Alloys and their Constituents..................................................8vo, | 2 | 50 |
| Wood's (De V.) Elements of Analytical Mechanics......................8vo, | 3 | 00 |
|    Treatise on the Resistance of Materials and an Appendix on the Preservation of Timber........................................8vo, | 2 | 00 |
| Wood's (M. P.) Rustless Coatings: Corrosion and Electrolysis of Iron and Steel.......................................................8vo, | 4 | 00 |

## STEAM-ENGINES AND BOILERS.

| | | |
|---|---|---|
| Berry's Temperature-entropy Diagram............................12mo, | 1 | 25 |
| Carnot's Reflections on the Motive Power of Heat. (Thurston.)......12mo, | 1 | 50 |
| Chase's Art of Pattern Making...................................12mo, | 2 | 50 |
| Creighton's Steam-engine and other Heat-motors....................8vo, | 5 | 00 |
| Dawson's "Engineering" and Electric Traction Pocket-book....16mo, mor., | 5 | 00 |
| Ford's Boiler Making for Boiler Makers............................18mo, | 1 | 00 |
| Goss's Locomotive Performance....................................8vo, | 5 | 00 |
| Hemenway's Indicator Practice and Steam-engine Economy...........12mo, | 2 | 00 |
| Hutton's Heat and Heat-engines...................................8vo, | 5 | 00 |
|    Mechanical Engineering of Power Plants......................8vo, | 5 | 00 |
| Kent's Steam boiler Economy......................................8vo, | 4 | 00 |
| Kneass's Practice and Theory of the Injector........................8vo, | 1 | 50 |
| MacCord's Slide-valves...........................................8vo, | 2 | 00 |
| Meyer's Modern Locomotive Construction..........................4to, | 10 | 00 |
| Moyer's Steam Turbines. (In Press.) | | |
| Peabody's Manual of the Steam-engine Indicator...................12mo, | 1 | 50 |
|    Tables of the Properties of Saturated Steam and Other Vapors.....8vo, | 1 | 00 |
|    Thermodynamics of the Steam-engine and Other Heat-engines......8vo, | 5 | 00 |
|    Valve-gears for Steam-engines................................8vo, | 2 | 50 |
| Peabody and Miller's Steam-boilers.................................8vo, | 4 | 00 |
| Pray's Twenty Years with the Indicator........................Large 8vo, | 2 | 50 |
| Pupin's Thermodynamics of Reversible Cycles in Gases and Saturated Vapors. (Osterberg.)................................................12mo, | 1 | 25 |
| Reagan's Locomotives: Simple, Compound, and Electric. New Edition. Large 12mo, | 3 | 50 |
| Sinclair's Locomotive Engine Running and Management.............12mo, | 2 | 00 |
| Smart's Handbook of Engineering Laboratory Practice...............12mo, | 2 | 50 |
| Snow's Steam-boiler Practice......................................8vo, | 3 | 00 |
| Spangler's Notes on Thermodynamics.............................12mo, | 1 | 00 |
|    Valve-gears....................................................8vo, | 2 | 50 |
| Spangler, Greene, and Marshall's Elements of Steam-engineering.......8vo, | 3 | 00 |
| Thomas's Steam-turbines..........................................8vo, | 4 | 00 |
| Thurston's Handbook of Engine and Boiler Trials, and the Use of the Indicator and the Prony Brake........................................8vo, | 5 | 00 |
|    Handy Tables...................................................8vo, | 1 | 50 |
|    Manual of Steam-boilers, their Designs, Construction, and Operation..8vo, | 5 | 00 |

Thurston's Manual of the Steam-engine......................2 vols., 8vo, 10 00
    Part I.  History, Structure, and Theory.....................8vo, 6 00
    Part II.  Design, Construction, and Operation................8vo, 6 00
    Stationary Steam-engines........................................8vo, 2 50
    Steam-boiler Explosions in Theory and in Practice............12mo, 1 50
Wehrenfenning's Analysis and Softening of Boiler Feed-water (Patterson) 8vo, 4 00
Weisbach's Heat, Steam, and Steam-engines. (Du Bois.).............8vo, 5 00
Whitham's Steam-engine Design....................................8vo, 5 00
Wood's Thermodynamics, Heat Motors, and Refrigerating Machines...8vo, 4 00

## MECHANICS PURE AND APPLIED.

Church's Mechanics of Engineering................................8vo, 6 00
    Notes and Examples in Mechanics..............................8vo, 2 00
Dana's Text-book of Elementary Mechanics for Colleges and Schools..12mo, 1 50
Du Bois's Elementary Principles of Mechanics:
    Vol.  I.  Kinematics...........................................8vo, 3 50
    Vol.  II.  Statics..............................................8vo, 4 00
    Mechanics of Engineering.  Vol. I......................Small 4to, 7 50
                              Vol. II........................Small 4to, 10 00
\* Greene's Structural Mechanics....................................8vo, 2 50
James's Kinematics of a Point and the Rational Mechanics of a Particle.
                                                                 Large 12mo, 2 00
\* Johnson's (W. W.) Theoretical Mechanics........................12mo, 3 00
Lanza's Applied Mechanics.........................................8vo, 7 50
\* Martin's Text Book on Mechanics, Vol. I, Statics.................12mo, 1 25
\*                            Vol. 2, Kinematics and Kinetics..12mo, 1 50
Maurer's Technical Mechanics......................................8vo, 4 00
\* Merriman's Elements of Mechanics..............................12mo, 1 00
    Mechanics of Materials.......................................8vo, 5 00
\* Michie's Elements of Analytical Mechanics.......................8vo, 4 00
Robinson's Principles of Mechanism................................8vo, 3 00
Sanborn's Mechanics Problems................................Large 12mo, 1 50
Schwamb and Merrill's Elements of Mechanism......................8vo, 3 00
Wood's Elements of Analytical Mechanics..........................8vo, 3 00
    Principles of Elementary Mechanics...........................12mo, 1 25

## MEDICAL.

Abderhalden's Physiological Chemistry in Thirty Lectures. (Hall and Defren).
    (In Press).
von Behring's Suppression of Tuberculosis. (Bolduan).............12mo, 1 00
\* Bolduan's Immune Sera.........................................12mo, 1 50
Davenport's Statistical Methods with Special Reference to Biological Varia-
    tions..................................................16mo, mor., 1 50
Ehrlich's Collected Studies on Immunity. (Bolduan.)...............8vo, 6 00
\* Fischer's Physiology of Alimentation..................Large 12mo, cloth, 2 00
de Fursac's Manual of Psychiatry. (Rosanoff and Collins.).....Large 12mo, 2 50
Hammarsten's Text-book on Physiological Chemistry. (Mandel.).......8vo, 4 00
Jackson's Directions for Laboratory Work in Physiological Chemistry...8vo, 1 25
Lassar-Cohn's Practical Urinary Analysis. (Lorenz.)...............12mo, 1 00
Mandel's Hand Book for the Bio-Chemical Laboratory..............12mo, 1 50
\* Pauli's Physical Chemistry in the Service of Medicine. (Fischer.)....12mo, 1 25
\* Pozzi-Escot's Toxins and Venoms and their Antibodies. (Cohn.)......12mo, 1 00
Rostoski's Serum Diagnosis. (Bolduan.).........................12mo, 1 00
Ruddiman's Incompatibilities in Prescriptions......................8vo, 2 00
    Whys in Pharmacy...........................................12mo, 1 00
Salkowski's Physiological and Pathological Chemistry. (Orndorff.)....8vo, 2 50
\* Satterlee's Outlines of Human Embryology......................12mo, 1 25
Smith's Lecture Notes on Chemistry for Dental Students.............8vo, 2 50

| | | |
|---|---|---|
| Steel's Treatise on the Diseases of the Dog...........8vo, | 3 | 50 |
| * Whipple's Typhoid Fever..................Large 12mo, | 3 | 00 |
| Woodhull's Notes on Military Hygiene ..............16mo, | 1 | 50 |
| * Personal Hygiene.........................12mo, | 1 | 00 |
| Worcester and Atkinson's Small Hospitals Establishment and Maintenance, and Suggestions for Hospital Architecture, with Plans for a Small Hospital.........................12mo, | 1 | 25 |

## METALLURGY.

| | | |
|---|---|---|
| Betts's Lead Refining by Electrolysis................8vo. | 4 | 00 |
| Bolland's Encyclopedia of Founding and Dictionary of Foundry Terms Used in the Practice of Moulding......................12mo, | 3 | 00 |
| Iron Founder..........................12mo, | 2 | 50 |
| " " Supplement......................12mo, | 2 | 50 |
| Douglas's Untechnical Addresses on Technical Subjects............12mo, | 1 | 00 |
| Goesel's Minerals and Metals: A Reference Book............16mo, mor. | 3 | 00 |
| * Iles's Lead-smelting........................12mo, | 2 | 50 |
| Keep's Cast Iron..........................8vo, | 2 | 50 |
| Le Chatelier's High-temperature Measurements. (Boudouard—Burgess.) 12mo, | 3 | 00 |
| Metcalf's Steel. A Manual for Steel-users..................12mo, | 2 | 00 |
| Miller's Cyanide Process.........................12mo | 1 | 00 |
| Minet's Production of Aluminum and its Industrial Use. (Waldo.)....12mo, | 2 | 50 |
| Robine and Lenglen's Cyanide Industry. (Le Clerc.).............8vo, | 4 | 00 |
| Ruer's Elements of Metallography. (Mathewson). (In Press.) | | |
| Smith's Materials of Machines.......................12mo, | 1 | 00 |
| Thurston's Materials of Engineering. In Three Parts...........8vo, | 8 | 00 |
| Part I. Non-metallic Materials of Engineering, see Civil Engineering, page 9. | | |
| Part II. Iron and Steel......................8vo, | 3 | 50 |
| Part III. A Treatise on Brasses, Bronzes, and Other Alloys and their Constituents....................8vo, | 2 | 50 |
| Ulke's Modern Electrolytic Copper Refining................8vo, | 3 | 00 |
| West's American Foundry Practice..................12mo, | 2 | 50 |
| Moulders Text Book.....................12mo, | 2 | 50 |
| Wilson's Chlorination Process....................12mo, | 1 | 50 |
| Cyanide Processes......................12mo, | 1 | 50 |

## MINERALOGY.

| | | |
|---|---|---|
| Barringer's Description of Minerals of Commercial Value. Oblong, morocco, | 2 | 50 |
| Boyd's Resources of Southwest Virginia.................8vo | 3 | 00 |
| Boyd's Map of Southwest Virginia...............Pocket-book form. | 2 | 00 |
| * Browning's Introduction to the Rarer Elements............8vo, | 1 | 50 |
| Brush's Manual of Determinative Mineralogy. (Penfield.)..........8vo, | 4 | 00 |
| Butler's Pocket Hand-Book of Minerals................16mo, mor. | 3 | 00 |
| Chester's Catalogue of Minerals...................8vo, paper, | 1 | 00 |
| Cloth, | 1 | 25 |
| Crane's Gold and Silver. (In Press.) | | |
| Dana's First Appendix to Dana's New "System of Mineralogy.."..Large 8vo, | 1 | 00 |
| Manual of Mineralogy and Petrography.............12mo | 2 | 00 |
| Minerals and How to Study Them................12mo. | 1 | 50 |
| System of Mineralogy...............Large 8vo, half leather, | 12 | 50 |
| Text-book of Mineralogy.....................8vo, | 4 | 00 |
| Douglas's Untechnical Addresses on Technical Subjects..........12mo, | 1 | 00 |
| Eakle's Mineral Tables......................8vo, | 1 | 25 |
| Stone and Clay Products Used in Engineering. (In Preparation). | | |
| Egleston's Catalogue of Minerals and Synonyms............8vo, | 2 | 50 |
| Goesel's Minerals and Metals: A Reference Book........16mo, mor. | 3 | 00 |
| Groth's Introduction to Chemical Crystallography (Marshall)....... 12mo, | 1 | 25 |

| | | |
|---|---|---|
| *Iddings's Rock Minerals......8vo, | 5 | 00 |
| Johannsen's Determination of Rock-forming Minerals in Thin Sections......8vo, | 4 | 00 |
| * Martin's Laboratory Guide to Qualitative Analysis with the Blowpipe.12mo, | | 60 |
| Merrill's Non-metallic Minerals: Their Occurrence and Uses..........8vo, | 4 | 00 |
| Stones for Building and Decoration......8vo, | 5 | 00 |
| * Penfield's Notes on Determinative Mineralogy and Record of Mineral Tests. 8vo, paper, | | 50 |
| Tables of Minerals, Including the Use of Minerals and Statistics of Domestic Production......8vo, | 1 | 00 |
| Pirsson's Rocks and Rock Minerals. (In Press.) | | |
| * Richards's Synopsis of Mineral Characters......12mo, mor. | 1 | 25 |
| * Ries's Clays: Their Occurrence, Properties, and Uses......8vo, | 5 | 00 |
| * Tillman's Text-book of Important Minerals and Rocks......8vo, | 2 | 00 |

## MINING.

| | | |
|---|---|---|
| * Beard's Mine Gases and Explosions......Large 12mo, | 3 | 00 |
| Boyd's Map of Southwest Virginia......Pocket-book form, | 2 | 00 |
| Resources of Southwest Virginia......8vo, | 3 | 00 |
| Crane's Gold and Silver. (In Press.) | | |
| Douglas's Untechnical Addresses on Technical Subjects......12mo, | 1 | 00 |
| Eissler's Modern High Explosives......8vo. | 4 | 00 |
| Goesel's Minerals and Metals: A Reference Book......16mo, mor. | 3 | 00 |
| Ihlseng's Manual of Mining......8vo, | 5 | 00 |
| * Iles's Lead-smelting......12mo, | 2 | 50 |
| Miller's Cyanide Process......12mo, | 1 | 00 |
| O'Driscoll's Notes on the Treatment of Gold Ores......8vo, | 2 | 00 |
| Peele's Compressed Air Plant for Mines. (In Press.) | | |
| Riemer's Shaft Sinking Under Difficult Conditions. (Corning and Peele)...8vo, | 3 | 00 |
| Robine and Lenglen's Cyanide Industry. (Le Clerc.)......8vo, | 4 | 00 |
| * Weaver's Military Explosives......8vo, | 3 | 00 |
| Wilson's Chlorination Process......12mo, | 1 | 50 |
| Cyanide Processes......12mo, | 1 | 50 |
| Hydraulic and Placer Mining. 2d edition, rewritten......12mo, | 2 | 50 |
| Treatise on Practical and Theoretical Mine Ventilation......12mo, | 1 | 25 |

## SANITARY SCIENCE.

| | | |
|---|---|---|
| Association of State and National Food and Dairy Departments, Hartford Meeting, 1906......8vo, | 3 | 00 |
| Jamestown Meeting, 1907......8vo, | 3 | 00 |
| * Bashore's Outlines of Practical Sanitation......12mo, | 1 | 25 |
| Sanitation of a Country House......12mo, | 1 | 00 |
| Sanitation of Recreation Camps and Parks......12mo, | 1 | 00 |
| Folwell's Sewerage. (Designing, Construction, and Maintenance.)......8vo, | 3 | 00 |
| Water-supply Engineering......8vo, | 4 | 00 |
| Fowler's Sewage Works Analyses......12mo, | 2 | 00 |
| Fuertes's Water-filtration Works......12mo, | 2 | 50 |
| Water and Public Health......12mo, | 1 | 50 |
| Gerhard's Guide to Sanitary House-inspection......16mo, | 1 | 00 |
| * Modern Baths and Bath Houses......8vo, | 3 | 00 |
| Sanitation of Public Buildings......12mo, | 1 | 50 |
| Hazen's Clean Water and How to Get It......Large 12mo, | 1 | 50 |
| Filtration of Public Water-supplies......8vo, | 3 | 00 |
| Kinnicut, Winslow and Pratt's Purification of Sewage. (In Press.) | | |
| Leach's Inspection and Analysis of Food with Special Reference to State Control......8vo, | 7 | 00 |
| Mason's Examination of Water. (Chemical and Bacteriological)......12mo, | 1 | 25 |
| Water-supply. (Considered principally from a Sanitary Standpoint)..8vo, | 4 | 00 |

| | | |
|---|---|---|
| * Merriman's Elements of Sanitary Engineering................8vo, | 2 | 00 |
| Ogden's Sewer Design.............................12mo, | 2 | 00 |
| Parsons's Disposal of Municipal Refuse....................8vo, | 2 | 00 |
| Prescott and Winslow's Elements of Water Bacteriology, with Special Reference to Sanitary Water Analysis............12mo, | 1 | 50 |
| * Price's Handbook on Sanitation...................12mo, | 1 | 50 |
| Richards's Cost of Food. A Study in Dietaries...............12mo, | 1 | 00 |
| Cost of Living as Modified by Sanitary Science...........12mo, | 1 | 00 |
| Cost of Shelter.............................12mo, | 1 | 00 |
| * Richards and Williams's Dietary Computer...................8vo, | 1 | 50 |
| Richards and Woodman's Air, Water, and Food from a Sanitary Standpoint...............................8vo, | 2 | 00 |
| Rideal's Disinfection and the Preservation of Food................8vo, | 4 | 00 |
| Sewage and Bacterial Purification of Sewage..............8vo, | 4 | 00 |
| Soper's Air and Ventilation of Subways. (In Press.) | | |
| Turneaure and Russell's Public Water-supplies................8vo, | 5 | 00 |
| Venable's Garbage Crematories in America................8vo, | 2 | 00 |
| Method and Devices for Bacterial Treatment of Sewage........8vo, | 3 | 00 |
| Ward and Whipple's Freshwater Biology. (In Press.) | | |
| Whipple's Microscopy of Drinking-water.....................8vo, | 3 | 50 |
| * Typhod Fever...........................Large 12mo, | 3 | 00 |
| Value of Pure Water......................Large 12mo, | 1 | 00 |
| Winton's Microscopy of Vegetable Foods.................8vo, | 7 | 50 |

## MISCELLANEOUS.

| | | |
|---|---|---|
| Emmons's Geological Guide-book of the Rocky Mountain Excursion of the International Congress of Geologists.............Large 8vo, | 1 | 50 |
| Ferrel's Popular Treatise on the Winds....................8vo, | 4 | 00 |
| Fitzgerald's Boston Machinist........................18mo, | 1 | 00 |
| Gannett's Statistical Abstract of the World.................24mo, | | 75 |
| Haines's American Railway Management..................12mo, | 2 | 50 |
| * Hanusek's The Microscopy of Technical Products. (Winton)......8vo, | 5 | 00 |
| Ricketts's History of Rensselaer Polytechnic Institute 1824–1894. Large 12mo, | 3 | 00 |
| Rotherham's Emphasized New Testament..............Large 8vo, | 2 | 00 |
| Standage's Decoration of Wood, Glass, Metal, etc............12mo, | 2 | 00 |
| Thome's Structural and Physiological Botany. (Bennett).......16mo, | 2 | 25 |
| Westermaier's Compendium of General Botany. (Schneider).......8vo, | 2 | 00 |
| Winslow's Elements of Applied Microscopy..................12mo, | 1 | 50 |

## HEBREW AND CHALDEE TEXT-BOOKS.

| | | |
|---|---|---|
| Green's Elementary Hebrew Grammar.....................12mo, | 1 | 25 |
| Gesenius's Hebrew and Chaldee Lexicon to the Old Testament Scriptures. (Tregelles.)....................Small 4to, half morocco, | 5 | 00 |

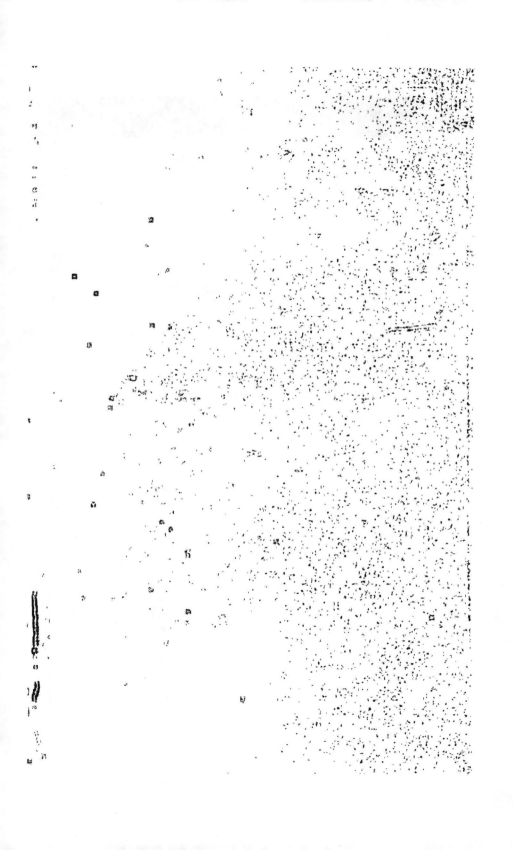

CPSIA information can be obtained
at www.ICGtesting.com
Printed in the USA
LVHW082243130622
721209LV00011B/721